Emerging Energy Technologies: Impacts and Policy Implications

To Jonathan Stern, an inspirational Head, with best wishes for the
future from the Energy and Environmental Programme

Emerging Energy Technologies: Impacts and Policy Implications

Michael Grubb, John Walker
Russell Buxton, Ted Glenny, Horace Herring, Bob Hill,
Claire Holman, Walter C Patterson, John Procter, Keith Rouse

The Royal Institute of International Affairs
Dartmouth

A CIP catalogue record to this book is available from the British Library and the US Library of Congress

Published by Dartmouth Publishing Company Limited, Gower House, Croft Road, Aldershot, Hants, GU11 3HR, England

Dartmouth Publishing Company Limited, distributed in the United States by Ashgate Publishing Company, Old Post Road, Brookfield, Vermont 05036, USA

√ISBN 1 85521 180 7

Cover by Twenty Twenty Design

Reprinted 1993

Printed and Bound in Great Britain by
Athenaeum Press Ltd, Newcastle upon Tyne.

Contents

Acknowledgments   xiii
About the Authors   xiv
Foreword   xv
Summary and Conclusions   xvii

PART I: HISTORICAL PATTERNS AND CURRENT CONTEXT

Chapter 1. Introduction and historical overview   **1**
1.1   Patterns of energy technology development   2
1.2   Technology diffusion rates and processes: an overview   6
1.3   Energy market and technology predictions   9

Chapter 2. The Evolving Pressures   **15**
2.1   Energy market trends   15
2.2   The economic context   18
2.3   Social and environmental constraints   20
2.4   External costs   23

Chapter 3. The Technological Menu   **29**
3.1   End-use technologies   30
3.2   Fossil fuel production and conversion   31
3.3   Nuclear technologies   33
3.4   Geothermal and renewable energy technologies   34
3.5   Criteria for case studies   36

PART II: DEMAND-SIDE TECHNOLOGIES

Chapter 4. Clean and Efficient Cars   43
4.1   Cleaning up emissions: end-of-pipe and alternative fuels   46
4.2   Trends in vehicle efficiency   48
4.3   Technical options for improving efficiency   49
    Engine improvements   52
    Vehicle and transmission design improvements   53
4.4   Prototype vehicles   54
4.5   Fuel efficient production vehicles   55
4.6   Trade-offs: efficiency, emissions, performance, and costs   56
4.7   Market barriers   58
4.8   Policy options   60
    Producer regulation and incentives   60
    Consumer incentives   62
    Regional issues   63
4.9   Prospects   65
4.10   Conclusions   66

Chapter 5. Energy Savings in Domestic Electrical Appliances   69
5.1   Electricity use in domestic appliances   70
5.2   Refrigeration efficiency   72
5.3   Performance variation and cost: an illustration from the UK   73
5.4   Appliance standards   77
5.5   Potential for energy savings   79
5.6   Energy labelling: the UK debate   79
5.7   Effect on manufacturers   82
5.8   Other white goods   83
5.9   Relevance to developing countries   84
5.10   Conclusions   84

Chapter 6. Energy Efficient Lighting Technologies   87
6.1   Energy use for lighting   88
6.2   Saving energy in lighting   89
6.3   Principal technologies   91
6.4   Compact fluorescent lamps   93
6.5   High frequency lighting   95
6.6   Need-control of lighting   97

6.7	Principal market areas	99
6.8	Pressures for increased uptake	100
6.9	Constraints on uptake	100
6.10	Potential impact	101
6.11	Policy mechanisms and options	102
6.12	Conclusions	104

Chapter 7. Building Energy Management | | **107**
7.1	The service sector	109
7.2	Building energy use and options	111
7.3	Building energy management systems	114
7.4	The history of BEMS	116
7.5	European market size and structure	117
7.6	Implementation of BEMS	118
7.7	Future developments in BEMS	120
7.8	Constraints and policy options	121
7.9	Conclusions	124

PART III: SUPPLY-SIDE TECHNOLOGIES

Chapter 8. Gas Turbine Systems | | **127**
8.1	Gas turbine technology	128
8.2	Performance and efficiency improvements	130
8.3	CCGT costs and economics	134
8.4	The availability of natural gas	137
8.5	Environmental issues	138
8.6	Electricity market developments	140
8.7	The market for gas turbine plant	141
8.8	Alternative fuels	143
8.9	Policy issues	144
8.10	Conclusions	145

Chapter 9. Clean Coal | | **147**
9.1	Coal-use technologies	149
9.2	Primary market areas	153
9.3	History	155
9.4	Pressures for increased uptake	157
9.5	Constraints on uptake	159

9.6   Policy mechanisms and options   160
9.7   Likely and potential impact   160

Chapter 10. Wind Energy   **163**
10.1   Characteristics of wind energy   165
10.2   Wind energy developments   169
10.3   Economics of wind energy   175
10.4   Resources and market potential   180
10.5   Market obstacles and current trends   184
10.6   Policy options   187
10.7   Conclusions   190

Chapter 11. Solar electricity from photovoltaics   **193**
11.1   Characteristics of the technology   194
11.2   Resource characteristics and environmental impact   196
11.3   Applications   197
11.4   Market size and development   200
11.5   Photovoltaic technologies: past, present and future   203
11.6   Policy issues in the application of photovoltaics technology   205
11.7   Future impact of photovoltaics   211
11.8   Conclusions   212
Box   Photovoltaic Technologies   209

PART IV: SYNTHESIS

Chapter 12. Case Study Comparison and Impacts   **215**
12.1   Common themes   215
12.2   Differences   218
12.3   Short and medium-term impacts   222
12.4   Longer-term implications   227

Chapter 13. The Future Context   **231**
13.1   The economic context   231
13.2   The environmental imperative   235
13.3   Technological implications   236

Chapter 14. Emerging Energy Policies   **239**
14.1   Research and development   239

14.2 Incorporating external costs   243
14.3 Supply-side technology policies   244
14.4 Promoting energy efficient technologies   245
14.5 Utility regulation and demand-side management   246
14.6 Technology transfer and developing countries   248
14.7 Globalization and competitiveness   250

Tables

2.1   Gross external costs of electricity generation in (West)
      Germany   24
2.2   Estimated environmental external costs associated with different
      power plants in the US   25
4.1   Technologies available for improved car fuel economy   51
4.2   Fuel efficient production cars on the UK market   55
5.1   Capacity, consumption and efficiency of UK refrigeration   75
5.2   Annual consumption UK 'best' refrigeration relative to US
      standards   78
5.3   Estimated energy consumption for refrigeration   80
6.1   Growth of lighting in developed and developing countries   89
6.2   Energy for lighting in the UK   90
6.3   Share of lighting energy in developed countries and lamp
      efficacies   91
6.4   Comparison of filament lamp with compact fluorescent
      lamps   94
6.5   Comparison of conventional fluorescent lighting and high
      frequency lighting   97
7.1   Estimated number and area of premises for the 12 main sub-
      sectors in service sector (1985)   110
7.2   Estimated expenditure on energy in service sector by enduse in
      UK   112
8.1   Gas turbine emission figures (g/kWh)   139
9.1   Comparative performance of coal-use technologies
      for electricity generation: indicative estimates   151
10.1 Complete costs of Danish windfarms installed in late 1980s   176
10.2 Costs of wind-generated electricity from currently available
      technology   177
10.3 US wind energy cost trends and projections   179
11.1 Projection of PV prices and sales   203

11.2 Technical PVgoals of the US government and industry for
      the year 2000   206
11.3 Research paths in the photovoltaic programme   207
11.4 Materials and efficiencies of PV devices   208
12.1 Assumed energy intensities for case study technologies used to
      illustrate potential impacts   226

Figures

4.1   Fuel consumption trends to 1988   50
4.2   Model-based calculation of the impact of improving efficiency,
      for a given base vehicle characteristics   58
5.1   Estimated consumption and running costs for UK white
      goods   71
5.2   Efficiency of refrigeration available in the UK   76
6.1   Improvement in lamp efficacy   92
6.2   Compact fluorescent lamps   93
6.3   Increase in lamp efficacy at high frequency   96
6.4   Electricity saving with 'Daylink' system   98
8.1   Operation of a gas turbine   129
8.2   Alternative gas turbine systems   132
8.3   Typical cost breakdown for CCGT and conventional sources 135
8.4   Fossil fuel and biomass generation costs, and impact of carbon
      tax   136
10.1 Modern commercial wind turbine   166
10.2 Measurements of wind energy developments in California 172-3
10.3 Distribution of wind energy generation   174
10.4 Distribution of European wind energy   183
11.1 Annual global PV module production   201
11.2 Market shares of PV sales   202
11.3 History of solar cell efficiencies   204
11.4 History of PV module costs   205
12.1 Breakdown of UK energy use by fuel and end-use sector and
      sectoral CO_2 emissions   224
12.2 Medium-term sectoral projections for the UK with potential case
      study savings   225
14.1 Energy R&D expenditure by IEA governments   240

Boxes

Demand-side case studies: Specific conclusions   xix
Supply-side case studies: Specific conclusions   xx
Characteristics of photovoltaic technologies   209

Research by the Energy and Environmental Programme is supported by generous contributions of finance and professional advice from the following organisations:

Amerada Hess • Arthur D Little • Ashland Oil • British Coal
British Gas • British Nuclear Fuels • British Petroleum • Caltex
Chubu Electric Power Co • Department of Energy
Department of the Environment
Eastern Electricity • East Midlands Electricity • ELF(UK)
Enterprise Oil • Exxon • GISPRI • Golden Rule Foundation • Idemitsu
Japan National Oil Corporation • Kuwait Petroleum • LASMO
Mitsubishi Research Institute • Mobil • National Grid
National Power • Neste • Nuclear Electric • Petronal • PowerGen
Sedgwick Energy • Shell • Statoil • Texaco
Tokyo Electric Power Co • Total • UK Atomic Energy Authority

Acknowledgments

Like any multi-author work, this book reflects the efforts of many people over a long period. We are particularly indebted to the authors of the various case studies who worked hard on draft manuscripts and upon bringing them to fruition, and accepted patiently the delays entailed in bringing the whole study together. All the case studies have also benefited from detailed comments provided by those who attended study groups on the project, and specialists who read carefully the relevant case studies - people too numerous to mention individually but all much appreciated. Special thanks are due to Ian Smart who read all of the draft Parts 1 and 4, and who provided many helpful insights.

The production owes much to the efforts of all the staff at the Energy and Environmental Programme. Jonathan Stern launched the project by pointing to the importance of studying the impact of technology development from a policy perspective, and he and Silvan Robinson as always helped to guide the study from initial stages to completion. Nicola Steen coordinated case study authors and review groups, and critically read the full manuscript. Matthew Tickle managed all the final stages of coordination and production of the volume, including graphics and typesetting, and John Milne created diagrams effectively and promptly. Rosina Pullman edited the full text, and handled the many queries from readers and authors. Finally, we are grateful to our families who have inevitably borne some of the extra pressure entailed in bringing this project to successful completion.

Michael Grubb and John Walker

About the Authors

Dr Michael Grubb is a Senior Research Fellow at the Energy and Environmental Programme of the Royal Institute of International Affairs. He directed the Institute's international programme of research on the energy, economic and political aspects of climate change, with a range of publications culminating in the two-volume international study, *Energy Policies and the Greenhouse Effect*. Prior to joining the Institute, he was at the Cavendish Laboratory in Cambridge and at Imperial College in London, where his studies resulted in a wide range of publications on the planning of electricity systems and the economic prospects for renewable energy sources.

John Walker is an independent consultant on alternative energy and conservation, and an expert adviser to the European Commission. He is also a Fellowship of Engineering Visiting Professor in the Principles of Engineering Design at the University of Southampton. Formerly, he worked for the Central Electricity Generating Board and National Power, specializing in alternative energies since 1979.

Other authors

Russell Buxton   Strategic Planning Manager
                          Rolls Royce Industrial and Marine Turbines Ltd
Ted Glenny   Technical Manager, Philips Lighting
Horace Herring   Energy Analyst, Business Planning, PowerGen
Bob Hill   Director of the Photovoltaics Applications Centre
                                      Newcastle upon Tyne Polytechnic
Claire Holman   Environmental Consultant
Walter C Patterson   Associate Fellow, Energy and Environmental
                    Programme, Royal Institute of International Affairs
John Procter   Lighting Consultant
Keith Rouse   Business Development Director, Molynex Holdings Plc

Foreword

One of the occupational hazards of running a research programme is the inability to predict how a study will evolve. It was not obvious that attempting to describe some important technologies and commenting on their policy relevance would result in a fourteen chapter book. The results are more intellectually rewarding, but have required a great deal more effort on the part of authors and programme staff, than anticipated.

Part of the difficulty has been finding suitable authors for the case studies which are intended to provide a current assessment of technologies, in language accessible to a lay person. For a variety of reasons, case studies fell by the way-side and were substituted and supplemented along the way. Particular gratitude is therefore due to case study authors who have revised and updated their studies as the timetable of the complete work slipped; and authors who were co-opted in the later stages of the study and required to meet tight deadlines in order to prevent further slippage.

Much of that slippage has been due to the priority given to Michael Grubb's work on the greenhouse effect. As the lead author of Parts 1 and 4, the manuscript was held up - but also benefited greatly from - the two volume study, *Energy Policies and the Greenhouse Effect*. Because of the substantial call on Michael's time as principal author and coordinator of the study, we are immensely grateful to John Walker for his assistance particularly on the case studies. John's experience and expertise on a range of energy technologies have been invaluable.

Too often, studies of technology are accessible only to those with a strong scientific and technological background. It is hoped that this study to some extent redresses that balance and places both the general subject, and specific technologies, within a policy context.

Jonathan Stern, March 1992.

Summary and Conclusions

Energy technologies shape the energy business and its impacts on society. Social, economic and environmental pressures and policies in turn determine the progress of different technologies. Yet the processes by which technologies develop and diffuse into markets are diverse and complex, and energy forecasts have often proved wrong in part because they have overlooked or misjudged the impact of technological change.

Innovation remains a largely unpredictable process, but past experience reveals that the selection and evolution of technologies, and the rate at which they penetrate markets, depends on: the scale and profitability of the technology; the extent of supporting infrastructure in place; and various identifiable economic and non-economic factors associated with the evolution of social and market trends and environmental pressures.

Market trends which affect energy developments include the shorter payback requirements arising from widespread capital scarcity and utility liberalisation, depressed but potentially viable fuel prices, expansion of gas resources and infrastructure, and the hiatus of nuclear power. Siting constraints and a range of environmental concerns from urban air quality to global climate change are assuming ever greater importance. Incorporating such 'external costs' may substantially alter relative costs of different options.

Many technologies could alleviate these pressures. Some are already quite well developed but have as yet made little impact on energy markets. Case studies revealed specific conclusions summarized in the box below. In general, the demand-side technologies considered (see Boxes) could have considerable impact over a 10-15 year period; in the UK, their rapid adoption would probably stabilize primary energy demand. Of the supply-side technologies considered (see Box), combined cycle gas turbines (CCGTs) will have the most impact over the next decade, though wind and coal fluidized beds could be significant. In the longer term, combined with continuing demand-side developments, these and other technologies could radically change the outlook for energy supply and demand.

But the progress and impact of such technologies will depend upon government and industrial policy. The take up of more efficient technologies will depend upon the extent to which governments are active in shaping consumer markets. Providing public information has an important role but not a dominant one. Beyond this, appropriate mechanisms vary greatly according to the technology and sector involved, and progress will be affected by the degree of policy sophistication in matching incentive structures to the market situation.

Major government R&D programmes have not focussed on these emerging technologies. But support for the initial diffusion of emerging supply technologies is important for all except CCGTs, and helps to drive company R&D; progress of all the supply technologies will also depend heavily on the extent to which environmental and other policies reflect the external impacts of conventional sources, and either raise their costs accordingly or subsidize cleaner technologies.

With half of primary energy consumption channelled through gas and electricity networks which are natural monopolies, utility regulation is important. Access to systems combined with procedures to ensure fair tariffs and connection charges are critical, especially for the small-scale electricity-producing renewable technologies. But trends towards liberalization address only one end of the system. The use of better metering technologies and development of regulatory systems which encourage demand-side investments offer powerful mechanisms for promoting energy efficient technologies.

The globalization of business makes competitiveness an issue of strategic importance in assessing emerging technologies. The potential benefits from emerging technologies are particularly large in the developing world, but uptake will lag perhaps by decades in the absence of international action including additional capital finance.

Though the timing is uncertain, the technologies examined all have promising futures because they can help ease the inevitably growing environmental and resource pressures. The countries and industries which will benefit most are likely to be those that anticipate this, and adopt measures which encourage the production and early adoption of emerging energy technologies.

Demand-Side Case Studies: Specific Conclusions

Average new car fuel efficiencies could be increased by 20-40% over the next 10-15 years without compromising safety, increasing life-cycle costs or reducing performance below current levels. Such improvements will not occur without government involvement, and will be offset by continuing traffic growth and any further increases in vehicle size and power. Effective policy measures could include fleet average efficiency standards (perhaps tradeable between manufacturers) combined with fiscal incentives to encourage manufacturers to make, and consumers to buy, more efficient and economical vehicles.

Domestic appliances offer great proportionate efficiency gains, without significant trade-offs against either cost or performance. Public information, including mandatory labelling, would be an important step though additional measures such as minimum efficiency standards and regulatory incentives could be required to exploit most of the potential.

New lighting technologies could reduce the overall energy intensity of lighting by 30-40% over the next decade, with further savings available, but the capital costs of some of these are much higher than current practice. Beyond these and a variety of other specific technical options, the overall management of energy in larger buildings provide further opportunities. Obstacles to efficiency improvements include ignorance, lack of industry standards, shortage of trained staff, lack of finance in some sectors, building leasing structures and accounting practices.

Uptake of these and other options would be increased by: public education of home-owners and building managers; established codes of practice in building design and management, perhaps including requirements for best practice or other minimum standards implemented through expanded building regulations; national programmes for public sector buildings; various forms of financing packages; and other regulatory approaches especially through utilities.

Supply-side Case Studies: Specific Conclusions

The use of gas-fired turbines in combined cycle plants (CCGTs) is now a fully commercial and highly competitive technology for power generation in most areas with access to adequate natural gas. Increasing the role of private finance in power generation and environmental constraints both benefit these plants, and they will penetrate markets rapidly during the 1990s. Other gas turbine systems to boost performance further and exploit fuels such as coal and biomass are feasible. But over this period they will have little impact and may not be developed at all without government support, because of the financial risks and various institutional obstacles, combined with the focus of capital on combined cycle plants.

Coal gasification is particularly promising as a highly efficient and clean technology for using coal in the longer term. Other clean coal technologies which use fluidized bed combustion in various applications have already been commercially demonstrated. Competition from CCGTs will be intense but progress will also depend upon further development and demonstration, emissions constraints which make coal and electricity industries pursue these technologies, and upon finance for applications in coal-rich developing countries with little gas.

Wind energy has developed rapidly during the 1980s especially due to support in California and Denmark. In the best locations it is already commercially viable without support and continued improvements are projected. Solar electricity from photovoltaics has also developed rapidly; it is particularly suited to isolated applications at present but could start to emerge into bulk power markets during the 1990s particularly in systems with solar-induced peak demands, and perhaps as systems integrated into building surfaces. For both wind and solar, key issues affecting progress include the conditions of access to electricity systems and electricity tariffs, the extent to which the external costs of fossil fuels are reflected in pricing or credits for renewable sources, and capital support particularly for developing country applications. Planning procedures, public attitudes, resource surveys and education are also important for wind energy.

PART I: HISTORICAL PATTERNS AND CURRENT CONTEXT

Introduction and historical overview

*Technical progress is implicit in every economic forecast
and is the joker in every economist's model.[1]*

Energy technologies determine the options for meeting the energy demands of modern society: the costs involved; the nature and scale of the energy industries; their role within and impact upon society; and their environmental consequences. Energy technologies are in a continuous state of evolution, and on timescales of decades and centuries there have been fundamental technical changes which have radically altered the uses to which energy has been put and the nature of energy systems.

Yet the process of technological change, the factors which determine the adoption and penetration of competing energy technologies, and the possible implications of new technologies for future energy policy and markets remain poorly understood. Macroeconomic projections often either assume constant technology, or assume that technical development is an autonomous process which is largely independent of policy. Support for energy technologies appears often to have reflected a belief that success is primarily a function of the scale of research and development (R&D) expenditure, whilst general energy policy in many countries is increasingly guided by beliefs that liberalized energy markets will select optimal energy technologies. Energy forecasts have often proved badly wrong because they have overlooked or seriously misjudged the impact of technological change.

This book examines the issues raised by energy technologies which are already emerging or which could start to enter into energy markets by

1. D.O.Hall et al, 'Assessment to renewable and non-renewable energy resources group report', in McLaren and Skinner, *Resources and World Development*, John Wiley & Sons, 1987.

the year 2000: the impacts which they might have, the factors which will determine their progress, and the implications for energy policy and energy business.

To this end, the study is divided into four main parts. Part I sets the background by reviewing some of the historical trends and experience from technology studies and forecasting, examines the evolving trends and pressures in the energy business, and outlines the menu of energy technologies which are available or which have been proposed. Part II then presents case studies of four demand-side technologies which explore in depth the impact which these could have for energy demand in the relevant sectors, and the conditions and policies which will affect their progress. In Part III, four more case studies examine the development of emerging energy supply/conversion technologies and the issues they raise. Finally, Part IV reviews the case studies for the common themes and the differences between them; the impacts which these technologies might have in different timescales; and the policy issues which arise.

This chapter introduces the subject by examining historical experience and some of the lessons which can be drawn from it. The study opens with a review of general patterns in energy technology development. Some key points from the broader literature on technology development and diffusion are then summarized. The chapter closes with a review of energy forecasting and predictions, and some of the points which this reveals about the issues which need to be taken into account in studying energy technologies.

1.1 Patterns of energy technology development

For thousands of years, people have sought to develop and improve ways of harnessing the energy available in their surroundings to do useful work. Wood fires were used before the beginning of cultivation, and coal was discovered and burned during the Iron Age. The ancient Egyptians learned to use watermills; windmills and waterwheels were widespread by the late Middle Ages.

The development of steam technology, for turning the thermal energy of wood and coal into usable mechanical energy, was one of the keys to the industrial revolution. In turn, the pressures of industrial growth drove the progress of steam power as applied to railways and shipping, and the technology of mining to supply the new demands. In the mid - to late -

19th Century the discovery of oil brought a new fuel to the market, and its marriage with the internal combustion engine secured the dominant technology for transport in the developed countries in the 20th Century. At the end of the last century, the development of electricity, generated then and still today predominantly by the steam turbine, heralded another profound change which was rapidly exploited and directed by the new-found wealth of consumers in fast-growing economies. The gradual improvement of oil and later gas recovery and processing techniques brought access to new reserves as the applications grew ever broader and more refined.

The history of energy supply - wood, coal, oil and latterly electricity and gas - is paralleled by the continuous development in the way it is converted into a useful product or service. The open fire evolved to the compact coal-fired boiler, now supplanted in many regions by gas boilers. The internal combustion engine has been continuously refined, and has many variants. The original incandescent lightbulb improved steadily, and is now accompanied by a wide variety of other lighting technologies. The electric motor found applications ranging from home fridges, washing machines and vacuum cleaners to heavy industrial plant.

Some analysts have perceived deeper patterns in such developments. The economist Schumpeter identified each of the general 'long waves' of economic development (often known as Kondriateff Cycles) as resting in large part on particular generations of technologies.[2] Many of these represented important new applications of the conversion and use of energy: the industrial revolution drawing on basic steam and iron-smelting technologies; the mid-19th Century expansion based on steam railways and Bessemer steel; and the third cycle, from the turn of the century, reflecting the entrance of electricity, petrochemicals and the motor vehicle.

Technology development is not an autonomous process, but is intimately associated with the needs and pressures created by society. Energy market conditions, as fashioned by the scale and nature of demand and resource conditions in combination with available conversion technologies, have been very important; and developments in energy technology conversely have interacted with socio-economic

---

2. J.A.Schumpeter, *Business Cycles: A Theoretical, Historical and Statistical Analysis of the Capitalist Process*, McGraw Hill, New York and London, 1939.

trends. For example, in many societies increasing wealth and the progressive emancipation of women has both driven and drawn heavily upon a wide range of domestic appliances for reducing the domestic workload. Rising environmental concerns as societies develop have also done much to fashion and constrain the nature of viable energy technologies, as discussed in more detail below.

More specific examples of market-technology interactions abound. The 1973 oil price increase stimulated not only a range of developments in demand-side and alternative fuel technologies, but also major responses within the oil business. Deep-sea drilling technologies which many had considered impossible, or impossibly expensive, entered rapidly. They were indeed quite expensive, but when the OPEC cartel collapsed and oil prices fell during the 1980s, drilling rig technologies and procedures were developed and exploited which reduced costs to levels which had been considered infeasible before, and which kept most such operations viable.

This does not of course imply that technology development is infinitely adaptable. Underlying technical and scientific developments have been crucial in fashioning what can be done and in opening new avenues for production and use. Two centuries of development in steam technology drew heavily upon underlying advances in both thermodynamics and metallurgy, and in turn contributed to these sciences. Laboratory studies of the strange phenomena induced by rotating magnets opened up electricity as a wholly new and versatile energy carrier. Studies of the phenomena of vapour phase transitions and heat transfers enabled refrigeration, air conditioning, and heat pumps. The seeds of future technologies are laid in research laboratories; prediction of possible applications is obviously hard, but some components can be seen as they emerge in applied science long before they appear in commercial energy technologies.

However, it is also striking that in some cases, technologies which were largely developed or very easy to foresee made little impression on markets until conditions changed. For example, the combined cycle gas turbine described in Chapter 8 is conceptually a simple extension of technologies which were available at the end of World War II. But they were not exploited in this way until changes in utility legislation and financial circumstances altered the economic basis for power generation and liberalized routes to new applications, and the discovery and development of extensive gas resources eased concerns about gas shortages and changed its image as a scarce 'premium fuel'.

New energy technologies face immense inertia, even when they have major advantages. This is particularly striking for major changes in supply technologies. New fuels have generally taken at least 50 years to establish a dominant position in energy markets.[3] Other innovations, such as in production and conversion processes, can occur more quickly, but still generally involve large-scale changes in heavy plant spanning decades. Demand-side changes tend to be governed either by the penetration of new energy-using activities, or once the use is established, by the turnover of new stock plus a few years for 'retooling' of production facilities. Feasible penetration rates and the factors governing them are taken up more fully below and in the case studies.

Turning to the role of national policy, governments have often sought to champion particular energy technologies through large-scale R&D programmes, especially in pursuit of national security goals. Such efforts have rarely led to very successful products. Examples include most of the largest post-war energy technology programmes: synfuels (synthetic fuels from coal); fast breeder reactors; and fusion power. Yet governments cannot avoid taking some role in the energy business, and hence by implication in the energy technology business. The need for appropriate, carefully directed policies is one of the themes addressed in this book.

Social factors and constraints are very important. Irrespective of the merits or otherwise of arguments about the safety and risks of nuclear power, publics in many countries have been very distrustful of this technology, and this in itself has often had a decisive influence on its development, or lack of it. Local resistance to siting, and broader safety and environmental concerns, have increasingly influenced a range of other energy developments. Few major decisions in the energy business today can be taken without careful consideration of the likely social reactions and environmental implications.

A final observation is that prediction has often been poor. Major trends in both energy markets and technologies have often not been foreseen, or have been badly misjudged. This chapter concludes with a review of some particular forecasting errors and some of the factors which

---

3. A.Andere, W.Haefele, N.Nakicenovic, A.McDonald, *Energy in a Finite World*, Ballinger, Cambridge MA, 1981.

contributed to them. First, we examine more closely some of the lessons drawn from the study of technology diffusion processes.

1.2 Technology diffusion rates and processes: an overview

Invention and the initial stages of technology development remain a mysterious and unpredictable process. A rule of thumb in marketing is that 70% to 90% of all new products fail. Earlier in the process, it has been estimated that only one per cent of innovations succeed. Humanity's continuing drive for innovation has been described as 'the triumph of action over analysis'.[4]

Beyond the initial chaotic phase of home or laboratory design and initial development, research has pointed to many factors governing the pace at which the survivors penetrate markets. Mansfield[5] demonstrated two particularly important economic factors governing the pace of diffusion. The first is the profitability of the new product for both producers and adopters: not surprisingly, larger profit margins give a greater incentive to drop old habits and move to new processes. Second, the rate is inversely related to the scale of investment required: bigger outlays induce greater hesitance in moving to new processes, the process may be directly constrained by capital shortages, and processes requiring bigger investments also often have longer lifetimes, all of which contribute to inertia.

This suggests a relatively simple picture in which technology diffusion is driven primarily by economic incentives and constrained by economic obstacles. But in reality many other factors - economic, social and cultural - also affect diffusion. Thus, 'prices convey certain kinds of information and send certain signals, but their roles and causal functions from a diffusion perspective would seem to be less important, especially in the early phases of diffusion, than mainstream economic theory would say.'[6] A study of the cotton industry illustrated a case where a new process with a payback of less than a year and consequent very high rate of return, low investment outlay, and easy availability, was nevertheless

---

4. All citations drawn from Jesse H.Ausubel, 'Rat Race Dynamics and Crazy Companies: The Diffusion of Technologies and Social Behaviour', in Nebojsa Nakicenovic and Arnulf Grubler (eds), *Diffusion of Technologies and Social Behaviour*, Springer-Verlag, Berlin/Heidelberg, 1991, p.10.

5. E.Mansfield, *Industrial Research and Technological Innovation*, Norton, New York, 1968.

6. Ausubel, op.cit., p.6.

still not adopted after many years by many of the companies concerned. Freeman[7] cites this example, and gives a useful review of current understanding:

It is now possible to synthesize some of the main conclusions deriving both from the sociological and from the economics research. Predictably the economists put the main emphasis on profitability, scale of investment, and relative costs, while the sociological work stressed characteristics of change agents, of opinion leaders and other groups within the potential 'adopter population'. Rather less commonly, individual studies stressed the importance of technical characteristics of new products and processes, compatibility and acceptability of innovations within an existing environment or system, and the role of political lobbies and government policies...

In other words, technology diffusion is a very complex phenomenon, affected by a wide variety of social and institutional as well as economic factors. Studies of emerging technologies cannot simply look at costs and cost projections, but need to consider a far wider range of issues which might affect the attitudes of those that might produce or adopt the technology, and the factors which affect the rate at which they may do so.

The rate at which technologies diffuse varies widely according to the technology and processes involved. The review by Freeman (ibid) reports that:

Any innovation which diffuses through half of a potential adopter population, or affects more than two-thirds of the relevant output of a good or service in less than ten years has enjoyed a rapid rate of diffusion. Most examples of this kind are in fact essentially substitution process ... More typically it takes 10 to 30 years for the majority of potential adopters to invest or purchase ... with completely new products or systems a slow diffusion rate of more than 30 years for the majority of firms or households to adopt is quite usual.

7. C.Freeman, 'The Diffusion of Innovations - Microelectronic Technology', in Robin Roy and David Wield, *Product Design and Technical Innovation*, Open University Press, 1986, (extracted from 'Some economic implications of microelectronics', in B.Lundvall and P.R.Christensen (eds), *Technology and Employment*, Alborg University Press, Denmark, 1981).

Energy supply technologies require very large investments, and
particularly for those which require adaptation of whole systems,
diffusion is clearly bound to be slow: more than fifty years has been
required for the major transitions from wood to coal, and coal to oil, with
expansion at rates rarely above 10% a year at most. Yet the complexity
of the process and the importance of different stages need to be
appreciated before applying general 'rules of thumb'. When technologies
substitute for existing uses, their growth rates tend to be limited by the
rate at which existing stock is retired (though the availability of a very
attractive new technology may accelerate retirement, and conversely,
penetration may be much slower if the benefits are insufficient to tempt
actors away from continuing with more familiar investments). For
wholly new applications, many more factors constrain the penetration
rate, including: economic advantages and obstacles; social attitudes and
understanding; and the extent to which supporting systems and
infrastructure are required and the rate at which they can be developed.

In this context Grubler[8] emphasizes that technology diffusion is a
complex sequence of overlapping stages, in some of which expansion
can be rapid. Thus the number of cars registered in the US expanded at
a rate of about 30% annually until the 1920s, after which growth reduced
to around 5%. The first phase reflected substitution of horses by cars,
with people buying cars as their horses retired, for use on existing road
and track infrastructure. Gas turbines for jet aircraft and peak-load
applications in existing electricity supply systems similarly grew at
around 30% a year during the 1960s, until most new commercial aircraft
used them and most electricity systems had adequate peaking capacity;
after which, growth rates declined sharply.

Given the complexities it is not surprising that technology prediction
is difficult, even subsequent to the chaotic stage of innovation.
Nevertheless, the history of energy forecasting and its assumptions
regarding technological uptake is such as to deserve particular
consideration.

---

8. Arnulf Grubler, 'Diffusion: Long-term Patterns and Discontinuities', in Nakicenovic
and Grubler (eds), op.cit., p.455.

1.3 Energy market and technology predictions

'It is difficult to make forecasts, especially about the future.' It is an old saying, but energy seems to have been a field particularly prone to wildly inaccurate forecasts, often linked to technological myopia. The late-19th Century concern that exponential growth in transport demand would soon leave London knee-deep in horse manure is perhaps the most famous example. Less well known is that the technology destined to deliver London from this fate was examined by the Royal Commission on the Motor Car in 1908, which according to Collingridge judged that the most serious issue raised by this new machine would be the dust thrown up from untarred roads.[9]

More recent examples are legion. According to many predictions made just two decades previously, the 1990s would be the decade in which the world would be gripped by a global oil shortage as reserves approached exhaustion and prices skyrocketed. At the same time, some still believed that nuclear power would be 'too cheap to meter'. As late as 1976 the UK Atomic Energy Authority presented a forecast for the UK with over 100GW of nuclear power, nearly half of it from fast breeder reactors, in place by the year 2000;[10] the realized figure for total nuclear capacity will be under one-tenth of this. Optimistic (and probably inconsistent) visions extended to the coal industry, which expected steadily expanding markets as the backbone of British energy (especially after the oil price shock), rather than the painful, extended and accelerating decline which has occurred.

One underlying factor in both these examples was a general assumption then made about energy markets, namely that energy consumption would continue to grow in line with economic growth. In the event, UK primary energy demand in 1990 was still below the level seventeen years

9. D.Collingridge, *The Social Control of Technology*, Open University Press, Milton Keynes, 1980.
10. Royal Commission on Environmental Pollution, *Nuclear Power and the Environment*, Sixth Report, HMSO, September 1976, pp.178-9. Of this projection the Commission noted: 'The AEA programme (104GW in 2000) represents the maximum possible UK programme subject to these constraints [of uranium availability and plutonium formation], but we were surprised to discover that it was considerably smaller than that originally advanced in their evidence (149GW in 2000) and much smaller than that calculated as desirable in a companion document (210-285GW in 2000)'.

earlier,[11] and total primary energy demand in the OECD grew by just 5% between 1973 and 1986, though rapid increase resumed in the late 1980s in some (though not all) regions following the oil price collapse.[12]

Some of the general reasons for errors in energy demand predictions have been summarized elsewhere.[13] The consistent over-prediction of future demand made in projections in the 1960s and 1970s was partly due to over-estimation of GDP and a failure to foresee the extent of oil price rises, but many other factors were involved. Analysts underestimated the extent to which energy efficiency could and would improve in response to price rises, in part because little interest had been paid to the technologies used in energy consumption. Also there was little consideration of the saturation of some important energy uses, and a failure to take account of structural changes in the economy, especially the shift to less energy-intensive industries in most developed economies. These difficulties were in some cases compounded rather than helped by over-reliance on complex computer models which served to obscure rather than clarify the critical assumptions.

Price, and to some extent demand, projections have also proved to be too heavily influenced by short-term trends, and have tended to draw too heavily on aspects which could be readily quantified, rather than less measurable factors. For example, price projections often reflected figures for proven reserves and neglected the more uncertain estimates of resources awaiting discovery - though these have proved crucial. Technical developments have again been important in giving economic access to many of these resources. Another aspect is that projections frequently paid insufficient attention to possible social and political changes, which in practice did much to shape energy markets - from the OPEC cartel through to changing patterns of consumer preferences and the impact of environmental concerns upon the energy business.

As noted above, issues of technology development and diffusion are closely intertwined with developments in energy markets, so that failures to predict general market trends are a major source of difficulty in

11. Department of Energy *Digest of United Kingdom Energy Statistics*, HMSO, 1990, Table A1.
12. *BP Statistical Review of World Energy*, BP, London, 1991.
13. M.Grubb et al, *Energy Policies and the Greenhouse Effect, Volume II: Country Studies and Technical Options*, Dartmouth, Aldershot, 1991, Chapter 4.

predicting technological trends. Furthermore, technological trends have served to create or to amplify more general market changes, and with hindsight it appears that insufficient appreciation of the potential of many energy technologies to develop and adapt to market conditions has contributed to the errors in market forecasting. Conversely, errors in predicting market conditions account for some of the trouble with technology predictions. The overall response to the oil shocks, in both supply and demand, is an obvious illustration. It is something of a chicken and egg problem.

By its nature it is impossible to predict technology development precisely. But apart from the general difficulties in predicting markets and market responses, other factors can be identified.

One aspect is the effect of scale, in two very different guises. Unexpected difficulties in scaling up from small prototypes to full-scale massive plants are legion: the problems and costs of large-scale reliable engineering have often been underestimated. Reliability has often been at the heart of this scale issue. If a plant is very large, it is subject to new kinds of stresses and is often designed in a complex fashion to extract the maximum performance. Yet failures in one component can halt operation of the whole plant. The history of electricity generation provides many examples where the scale economies available on paper have been more than offset by such difficulties, so that forecasts of technological performance and progress have proved very optimistic. Each major scaling up in the size of generating sets has been accompanied by greater initial reliability problems. In the UK, a decade of patchy performance followed the step from 200MW to 500MW and 660MW sets, inducing scepticism that may have prevented plans for 1,300MW sets proceeding.

The point can be pushed too far. The 660MW sets did eventually mature into reliable systems which met their promise, generating power at substantially greater efficiency and lower cost than is possible from smaller units. Similarly, the North Sea oil rigs, with the combined resources of the multinational oil companies and government backing, have developed to extract oil in harsh conditions at lower costs than many thought possible. The point is more that big engineering requires major commitments and time to iron out often unforeseen problems. The experiences have helped to change the former climate of 'big is beautiful' into one of considerable scepticism about new departures in large-scale engineering.

Conversely however, technologies which can be produced en masse on production lines nearly always become cheaper as the scale of production grows. For products in the mechanical industry, a doubling of production volume (combined with the learning involved as an industry expands) has typically reduced costs by 15-25%; for photovoltaic cells the figure has been perhaps 30% to date.[14] Because of the drive for very large-scale supply plants this has been most evident on the demand side; but as the case studies below illustrate, it is also emerging as an increasingly important feature for supply technologies as the pursuit of continuing scale increases gives way to a focus on the greater reliability and shorter lead times offered by smaller units. Yet projections often use data for technologies which are already proven and perhaps established in niche markets, neglecting or underestimating the inevitable decline in costs with mass production.

Other issues in technology prediction can be tied directly to the observations made above about the factors which influence energy technology development. The impact of social and environmental constraints has often been underestimated and misjudged, as with nuclear power. Knowledge gained from the literature on diffusion rates and processes has rarely been applied critically to the prospects for energy technologies. This has contributed both to unrealistically high projections of growth rates for politically favoured technologies, and neglect of those which can more readily meet market needs and fill niches and thus can expand from obscurity relatively rapidly.

Predictions might have been better had more attention been paid to developments in underlying technologies and how they might affect energy technologies, though by its nature this is difficult. Perhaps more striking is the limited attention paid in mainstream energy projections to the possible impact of technologies which were already well beyond the drawing board, and in many cases already available at least in prototype forms. Whilst emphasizing the difficulties, it is at least plausible that more detailed attention to the technologies which were already available, their potential for improvement, and the factors which might affect how they emerge into energy markets could often help to form a better view

14. Haruki Tsuchiya, 'Photovoltaics cost analysis based on the learning curve', Proc. Solar World Congress 1989, Japan.

of possible future developments in energy markets. It is upon such 'emerging' technologies that this study concentrates.

Making better predictions is of more than academic interest. Immense resources have been wasted because of mistakes made in projecting future conditions and assessing (or failing to assess) technological options, and damaging tensions and conflicts have arisen from the clashes generated when big industrial programmes have failed to recognize important social and environmental constraints. A better understanding of likely future trends and conditions, the technologies available for meeting future needs, the options for harnessing those technologies, and the constraints upon them, would be of great value.

This chapter has illustrated the extent to which the process of technical change is complex and remains inadequately understood. The standard economic theory that the pace of penetration is determined by the profitability and scale of the investment required is complemented by experience which demonstrates the importance of a host of other factors. These include broad economic and social trends and environmental constraints, in addition to more specific issues associated with the characteristics of the existing and new markets potentially open to new technologies. A precondition for studying the potential impact of emerging technologies is thus an appreciation of existing and likely future trends, and the implications they may have for energy market conditions. The next chapter examines these evolving pressures.

The Evolving Pressures

The changing pressures in the energy business arise from many quarters. Some reflect internal changes in the nature of energy supply and demand. Others are driven by broader trends in economic circumstances and political attitudes towards the role of government in big business. Others arise from changed social and environmental concerns. The discussion in this chapter follows this categorization.

2.1 Energy market trends

Perhaps the biggest change in the energy business from a decade previously is the outlook for oil. When OPEC dramatically raised the price of oil in 1979 following the outbreak of the Iran/Iraq war, it was not widely perceived that this second sharp price rise carried the 'seeds of its own destruction', in that it accelerated the trend of diversifying away from OPEC oil by stimulating both moves into alternative fuels, and the exploration and exploitation of other producing regions. The OPEC cartel could not hold, and oil prices first slid and then crashed in 1986 before stabilizing in the region $15-25/bbl, little above the real oil price before the price rise of 1973.

Discoveries in the Middle East have added still more to proven reserves. Whereas global proven reserves in 1970 stood at 30 years production, in 1990 the figure was 44 years remaining production. The response to the Iraq conflict of 1990/91 demonstrated the extent of slack in the market: about 10% of annual global production was lost, but spare production capacity covered the loss with little difficulty, and throughout the conflict, price rises were more due to nervousness than any supply shortfall.

This nervousness is however an important feature of the oil market. The impact in 1990/91 might have been far greater had only a little more capacity been lost, and the 'overhang' of oil production capacity is likely to decline during the 1990s. The major producing regions of the Middle East and the former Soviet Union are far from stable politically. The evidence of the overall supply/demand balance suggests that prices could remain low during the 1990s, despite resumed growth in oil demand. But political storms amplified by market nervousness could nevertheless make them volatile, and most analysts expect conditions to tighten around the turn of the century, with a relative decline in non-OPEC production returning greater power over prices to OPEC.[1]

Another changing factor is the availability of gas. While proven reserves of coal and oil have grown steadily, those for gas have expanded dramatically; reserve estimates almost doubled during the 1980s, and now equate to about sixty years of gas production at current levels, with a total energy content approaching that of oil reserves. The estimated total resources more than doubled in the thirty years to 1990s, and are still more than twice the proven reserves. The geographical spread of gas resources also increased, to 85 countries by 1989 compared with about 40 in 1960,[2] though about 75% of proven reserves are in the former Soviet Union and Middle East. The development of gas pipelines, and systems for liquefying natural gas, have made gas available to much broader markets. As with oil, shorter-term variations in the balance between available supply and demand may lead to price fluctuations, but gas has clearly achieved an actual or potentially central role in the energy balance in many areas, and in most could retain that position for several decades.

A third important factor is the hiatus in nuclear power. The accumulation of public opposition and economic difficulties has meant that many developed countries have abandoned their visions of nuclear expansion, and even taking into account its remaining strongholds in France and Japan, nuclear construction in the OECD has all but ceased. The political changes in eastern Europe have brought plans for nuclear expansion there, already in trouble, into grave doubt; indeed, the capacity could decline as reactors are closed for safety reasons. Only in a few parts

---

1. For an extensive study of the factors affecting oil prices see J.Roeber, *Oil and Energy Markets: Formation of Prices*, RIIA, forthcoming, 1992.
2. World Energy Council, *1989 Survey of Energy Resources*, WEC, London, 1989.

of the industrializing world does nuclear power seem likely to expand significantly over the coming decade or two. That in itself is a huge reversal from the outlook twenty years previously. It means that the 1990s, the decade for which earlier projections pointed to the inevitability of a decisive move away from fossil fuels, the world is becoming more dependent on fossil resources than ever.

It is in part because of these factors, combined with the legacy of the oil shocks and the extent of environmental concerns discussed below, that another trend deserving specific mention has emerged, namely greater interest in the demand side of the energy equation, and the options for influencing it. The response to the oil shocks revealed that energy demand, especially in developed economies, is responsive to changing conditions. As the one option that simultaneously reduces all the problems associated with energy provision, steadily increasing interest and importance has been attached to the options for improving end-use energy efficiency, and they feature strongly in this book.

However, efficiency improvements will not offset the need for extensive new supply capacity during the 1990s and beyond. Global energy consumption during 1986-90 grew at an average rate of 2.1%, with growth in all the major fuels; excluding the former Soviet Union and eastern Europe, the figure is 2.7% and remained strong through the period of the Gulf crisis of 1990/91.[3] The pressures for growth in the developing world especially are so strong that increasing global energy consumption seems inevitable even if very strong measures to conserve energy are taken.

Within the overall energy mix, electricity production - which currently takes about 30% of primary energy production and a much higher proportion of energy investments - continues to grow faster than the demand for overall primary energy. In the period 1970-1985 electricity demand did not however grow nearly as rapidly as expected, and many developed countries suffered from considerable over-capacity. Construction of new plant almost halted in some countries in the 1980s. But much of this over-capacity has already been absorbed by demand growth, which in many regions accelerated again in the late 1980s. With many of the plants installed in the post-World War II boom having to be retired, extensive new electricity generating capacity is likely to be

---

3. Derived from *BP Statistical Review of World Energy*, BP, 1991.

required worldwide in the 1990s and beyond, irrespective of measures to improve the efficiency of electricity use.

2.2 The economic context

To the end of the 1960s, the dominant need in energy provision throughout the world was to meet the rapidly growing demand. To this end, government resources were often channelled through nationalized energy industries. Despite the difficulties encountered, the efforts were largely successful, as witnessed by the spectacular growth in energy supplies.

In the retrenchment that followed the oil price shocks, the appropriateness of this model was increasingly called into question. As economies adjusted to recession following the oil shocks, national construction programmes in many of the most developed economies charged on into extensive over-capacity, becoming increasingly adrift from the needs for more responsive and flexible developments. The pressures for change grew. In eastern Europe, shielded from the oil shocks by the system of central planning and pricing including heavy subsidies for Soviet energy, expansion continued unaltered, but the recent upheavals have opened the system to similar pressures. New plant will still be required in many regions as noted above, but only in the developing world is rapid expansion now the dominant industrial imperative.

In parallel, mainstream economic philosophy in many countries has swung decisively towards reducing the role of detailed government planning in major industries, with greater enthusiasm for more competitive market processes. In the field of energy, this is being felt most in the utility sectors. The combined privatization and liberalization of UK electricity at the end of the 1980s is an extreme example, but more commercial and private-market-based approaches towards utility operation are becoming increasingly widespread. These include the financial terms of utility planning, allowance for competition from independent suppliers, and in some cases 'common carrier' legislation (in both gas and electricity) which allow industrial consumers to negotiate directly with producers, with the utility as provider of transportation services.

This has several implications for emerging energy technologies. Private finance is very different from government finance. It requires a high rate of return and is much more averse to risks and delays. Projects which take 6-10 years to complete and which provide returns over the subsequent 20-40 years used to be the backbone of energy utility planning. Private finance generally considers such projects with extreme scepticism, especially if there are any alternatives available which promise to pay back most of the cost within a few years. The changes are less marked in the fossil fuel production business, which has always had a bigger role for private industry dealing in big investment commitments with big returns, but it has a profound impact on the electricity business where the choice is wider.

In developing countries and central and eastern Europe, different circumstances lead to similar or more extreme consequences for financial attitudes. Intense scarcity of capital results in a focus on projects with low capital costs and short lead times. Pressures on capital have indeed helped to raise interest rates worldwide. In general therefore, energy investments are being considered in a world in which capital is more scarce, interest rates are higher, and financial time horizons are shorter than in most of the post-war period. These circumstances are widely expected to persist for many years, and various implications of this are reflected in the case studies.

Another implication of the trend away from government intervention in many cases has been a decline in state support for uneconomic industries. This has affected coal particularly in regions such as Europe where direct and indirect subsidies have been reduced, and nuclear power. Obviously this tends to make these sources less competitive relative both to alternative fuels and imports, and tends to raise energy prices. Many east European countries and former USSR republics, which relied upon heavily subsidized oil and gas, are now required to pay near market prices in hard currency. The trend is not universal, however, and subsidies remain widespread particularly (though not exclusively) in some developing and east European countries, along with other forms of financial incentives such as the tax breaks for domestic oil production in the US National Energy Strategy.

A parallel development has been the growing scepticism towards big government R&D programmes, where, as noted in Chapter 1, the failures have often been more notable than the successes. In Europe especially there has also been a retreat from support for national 'champion' technological areas, because the national markets are not large enough to make such outlays worthwhile. However, in general there has been clear recognition of the need for some forms of industrial support, as illustrated by Japan, and in other areas partly as a response to the onslaught of Japanese high technology. There has thus been a trend away from national programmes for developing particular technologies towards more regional support for particular areas of leading-edge industrial activity, most notably information technology.[4] As yet, energy has not featured much in such developments. Some of the issues and options involved in such support are discussed in some of the case studies and in the concluding chapter.

These changes reflect another broad economic trend, namely that towards 'globalization'. Knowledge about technologies can transfer between regions ever more rapidly, and economic systems themselves are becoming steadily more interdependent. Energy systems have not been immune from the trend, and fluctuations in national energy intensities between different countries appear to have become more closely coupled over time.[5]

2.3 Social and environmental constraints

For many decades the main task of energy industries was to supply the energy demanded by industrializing societies in whatever way the industry could meet it. As the scale of the energy business has grown, and as people have become richer, that has changed. This is reflected in two convergent factors: the increasing scale of social and environmental impacts; and increasing sensitivity towards social and environmental issues more generally.

The repercussions have been most striking in respect of nuclear power, hydroelectric power, and general difficulties with siting energy facilities - the 'NIMBY' (not in my backyard) syndrome. In developed countries, gaining planning permission for a plant, the facilities for bringing fuel to it, and for transporting the power away, is a major hurdle - often the decisive one. Problems in gaining consent for the construction of electric transmission lines is increasingly claimed to be one of the major constraints on power system expansion and interconnection in the US.

4. M.Sharp and P.Holmes, 'Conclusions: Farewell to the National Champion', in M.Sharp and P.Holmes, (eds), *Strategies for New Technology*, Philip Allan, London, 1989.
5. J.Edmonds and J.A.Reilly, *Global Energy: Assessing the Future*, OUP, 1985, p.51.

Whilst such constraints are probably having the greatest impact on energy industries in OECD countries, others are not immune. Indeed, in central and eastern Europe environmental concerns about nuclear, hydro and coal were one of the focal points of opposition prior to the revolutions. Worldwide, opposition to hydro dams from those who would be forced to leave their homes has been reinforced by that from people concerned about the broader social and ecological consequences of large hydro schemes, and such opposition now forms the major constraint on large hydro developments. Multinational operations in developing countries have also come under fire both in the field and in the country of origin for their environmental impacts.

Other environmental issues have grown steadily in their impact, in a spiral of concern and response. The London smogs of the 1950s resulted in thousands of deaths, and led to widespread 'smokeless zones' where raw coal burning was banned. Fumes from cars became a serious concern in the 1960s and led to widespread legislation to clean up exhausts, first in Japan and the US, and later in Europe. In some areas these measures have in turn been rendered insufficient by continued growth in vehicles usage, as outlined in Chapter 4.

More regional concerns about energy emerged with Scandinavian claims that their river and lake ecosystems were dying as a result of acid deposition from power stations in other countries, followed by the phenomenon of forest dieback. In Europe and North America this emerged as a major issue in regional diplomacy. Intensive study demonstrated complex links with acid deposition and a range of other pollutants from power stations, cars, and various other energy-related sources, though the full mechanisms are still far from entirely understood. Legislation in both Europe and the US has now been passed to mandate substantial reductions in sulphur dioxide by large emitters such as power stations and oil refineries, at considerable cost and with a profound impact on coal markets where low-sulphur coal now usually carries a significant premium. Related concerns, on which legislation is less developed, include the production of NO_x, a precursor to nitric acid and urban smog. This is produced to varying degrees by all combustion systems, from power stations to cars, as the combustion oxidizes nitrogen already in the air.

Finally, the greenhouse effect has emerged into the realm of international politics. The process of turning hydrocarbon fuels into CO_2

and water is a major contributor to the build-up of gases which trap heat in the lower atmosphere. This is likely to lead to climatic changes of an uncertain but potentially profound nature and magnitude, probably with a rise in sea level and some potential for unpleasant global climate-related surprises. The scientific report of the Intergovernmental Panel on Climate Change,[6] set up to examine the issue, endorsed concerns. A global framework convention on climatic change is scheduled to be signed in June 1992, and this is likely to be but the first step on a long road towards effective international limitation of CO_2 emissions, along with regional initiatives such as the European Community target to stabilize CO_2 emissions and associated Commission proposals for energy/carbon taxation. The nature and implications of these various environmental pressures are examined more fully in the final chapter in the light of the case studies.

Environmental policy has evolved along with the specific concerns. Initial responses focused on particular technologies and uses, for example, smokeless zones and catalytic converters. Local planning enquiries were heavily used as ways of vetting local impacts, often in conjunction with Environmental Impact Assessments for big projects, a process since standardized and formalized in many countries.

The potential inefficiencies and other drawbacks of explicit technology-based legislation has led to other forms of control. Rather than stipulating particular technologies, the European Large Combustion Plant directive introduced a 'bubble' approach, in which countries had to meet targets for total sulphur reductions, by whatever means seemed most appropriate.[7] The US Clean Air Act of 1990 has gone one step further, mandating the overall sulphur reduction required but leaving utilities free to trade 'emission permits', so as to find the least costly way of meeting the emissions requirements.

The increasing costs of environmental control, dissatisfaction with traditional approaches, and the need for mechanisms which harness market forces to promote cleaner technologies, have raised more explicit questions about the costs of such 'external' impacts, and the extent to which these can and should be reflected directly in energy pricing.

6. J.T.Houghton, G.J.Jenkins and J.J.Ephrams (eds), *Climate Change: The IPCC Scientific Assessment*, CUP, Cambridge, 1990.
7. N.Haigh, 'EEC: Policy and Implementation', *European Environment Yearbook*, DocTers International, London, 1991.

Environmental impacts thus now form part of a broader and crucial debate about the external costs of energy provision.

2.4 External costs

'External costs' are the costs of a given activity which are not reflected in the market price. In energy, external costs include environmental impacts, but also include subsidies of various forms, including government R&D expenditure. Many argue that the long-term costs of depletion of finite resources are not adequately reflected in the market price, whilst in the US the costs of military forces associated with helping to secure reliable supplies from foreign sources have also been cited. Two extensive and widely publicized studies of external costs, concentrating on electricity production, are those of Hohmeyer[8] for West Germany, and Ottinger et al[9] for the US.

Hohmeyer's study adopted a broad 'top down' approach which aggregated the impacts of various pollutants from fossil fuels in power production (which for the German system, as with many, are dominated by coal), and considered various components of direct and indirect support and economic externalities. The broad range of uncertainty was reflected in the results, which are summarized in Table 2.1. The major factors are those of environmental impact and depletion surcharges (to reflect implicit additional costs of resource depletion), and R&D expenditure for nuclear. In all, Hohmeyer estimated total external costs to be in the range 1.35-3.3p/kWh for fossil fuels, and 3.62-7.3p/kWh for nuclear.

The Ottinger et al study focused explicitly on environmental impacts, without considering the various direct and indirect support or depletion charges, but with a much more disaggregated study of the impacts of different plant types. The study simply gives point estimates without uncertainty ranges for most components, with results summarized in Table 2.2. This indicates the extent to which external costs depend on different technologies and fuels. The point estimate for the external costs of conventional coal stations exceeds the upper bound of Hohmeyer's range of environmental costs, though nuclear is within it. A recent review

8. O.Hohmeyer, *Social Costs of Energy Consumption*, Springer-Verlag, Berlin 1988.
9. Richard Ottinger, David Wooley, Nicholas Robinson, David Hodas, Susan Babb, *The Environmental Costs of Electricity*, Oceana Publications, New York, 1990.

Table 2.1 Gross external costs of electricity generation in (West) Germany (1990p/kWh)

	Fossil fuels (mostly coal)	Nuclear
Environmental effects	0.42-2.28	0.45-4.49
Depletion surcharge	0.85	2.21-2.32
Goods and services publicly supplied	0.02	0.05
Monetary subsidies	0.11	0.06
Public R&D transfers	0.01	0.90
Total	1.35-3.30	3.62-7.73

Note: converted from 1982pf/kWh to 1990p/kWh at 1pf=0.38p. Numbers may not add up due to rounding.

Source: O.Hohmeyer, *Social Costs of Energy Consumption*, Springer-Verlag, Berlin, 1988.

and comparison of these and related studies of external costs is given by Lockwood.[10]

Estimates of externalities are inevitably uncertain. The range in Hohmeyer's estimates reflects this, but comparison with the Ottinger et al study and analysis of the components reveals still wider possible differences. Hohmeyer's estimates of depletion surcharges have been criticized as much too high; conversely, he uses much lower values of 'human costs' in terms of the monetary equivalent used for health damage and morbidity than Ottinger et al, which helps to account for the difference in environmental costs from fossil fuels (Hohmeyer used a 'statistical value of life' at about £330,000 compared with the figure of £2m used by Ottinger et al). Despite the wide range for nuclear environmental costs given by Hohmeyer, the real uncertainty is still

10. Ben Lockwood, *The Social Costs of Electricity Generation*, Centre for Social and Economic Research Into the Global Environment working paper GEC 92-10, Department of Economics, University College, London, forthcoming 1992.

Table 2.2 Estimated environmental external costs associated with different power plants in the US (1990p/kWh)

	Coal conventional	Coal IGCC	Gas CCGT	Nuclear PWR[a]
Human	2.41	0.61	0.19	-
Material	0.19	0.05	0.03	-
Visibility	0.33	0.06	0.05	-
Total	2.93	0.72	0.27	1.62
Climate[b]	0.94	0.94	0.55	

Notes: [a]damages associated with nuclear operation and accidents. This is dominated by the assumed risk of accident.[b]Climate impacts have been separated because of the immense uncertainties involved (see text).

Source: Ottinger et al *The Environmental Costs of Electricity*, Oceana Publications, New York, 1990 as presented by B.Lockwood, *The Social Costs of Electricity Generation*, Centre for Social and Economic Research Into the Global Environment working paper, Department of Economics, University College, London, forthcoming 1992.

larger; these estimates are dominated by the estimated risks of severe accidents, which vary by a factor of 100.

These uncertainties are however mild compared with those associated with attempts to cost the impact of climate change. Lockwood cites a number of studies which use different (and largely incompatible) approaches, including three global studies.[11] The estimated global costs associated with climatic change projected by about the middle of the next century in these three studies range from a few hundred billion dollars annually (a small fraction of projected global GNP) to numbers several times global GNP, for an estimate based on possible worldwide deaths of 1-10m people annually valued at £1-10m per life. National studies reveal a wider range still, with some studies for example

11. ibid.

concluding that the costs of climatic change in the US may be very small. The point estimates of the Ottinger et al study (Table 2.2) reflect an estimate of the costs of mitigation (absorbing CO_2 through reforestation) rather than potential climate impacts.

Estimates of external costs, especially environmental ones, are thus highly uncertain. Nevertheless, the central point remains that the few attempts that have been made have judged external costs to be significant in comparison with the costs that are reflected in energy markets. This obviously increases the value of energy conservation, and of sources which do not impose these costs. Hohmeyer estimated the external costs associated with production from the renewable energy sources of wind energy and photovoltaics, which include land use, health impacts, and R&D. Broader economic effects arising from balance-of-trade impacts and employment effects (including reduced social security payments) can be positive or negative; as compared with fossil sources, Hohmeyer estimates that these external economic effects would be positive for wind and PV. In all, taking the average external impacts of current electricity production in the former Federal Republic of Germany into account, Hohmeyer estimated net value of external impacts from these renewable sources to be:

Wind   -2.1 to -4.6 p/kWh

Photovoltaics   -2.6 to -6.4 p/kWh

The negative value signifies net benefits. Further research and debate will undoubtedly lead to revised estimates, but the central point that external factors are economically significant, and favour cleaner technologies and especially energy conservation and some renewable sources, seems unlikely to change.

There have already been some moves to reflect external costs explicitly in energy policy. They have been used as a principal justification for renewable energy subsidies in Denmark at a level of 0.03ECU/kWh (about 2p/kWh). In the US where some utilities are required to compare the costs of conservation against new energy supplies, a number of states have added credits of 10-15% to the nominal value of conservation to reflect the avoided external costs, and Germany has taken similar steps. Carbon tax proposals reflect an explicit attempt to reflect the external costs of CO_2 emissions, and recognition of external costs are also of course implicit in existing measures to mandate reductions of sulphur and other pollutants.

The trend towards recognizing and incorporating external costs to some degree thus forms another important evolving pressure in the energy business, which could significantly alter the relative economic standing of different options. The impact of these and other changes will depend upon the technologies which are available to exploit the changing conditions, to which we now turn.

The Technological Menu

Ideas for ways in which energy can be produced, converted, and put to use to meet human needs span a huge range. As a precursor to the more detailed case studies, this section notes the principal technologies which have been explored and which either are already starting to make a significant impact on energy markets, or which proponents have argued could do so. The final section lists and summarizes the key features of the technologies selected for detailed case studies in Parts II and III.

A recent critical review of most of the technologies outlined in this chapter has been given by one of the editors elsewhere.[1] Several more extensive technology studies and compendia have been published.[2] The following cursory review makes no attempt to duplicate such efforts. Rather, it is intended to give the reader an impression of the range of technological options which are being considered and their approximate status at present. Unless otherwise stated, further details on the technologies may be obtained from the studies cited in footnotes 1 and 2, and additional references are only given if they provide information on technologies not covered in these studies, or if they illustrate recent

1. M.Grubb et al, *Energy Policies and the Greenhouse Effect, Volume II: Country Studies and Technical Options*, Dartmouth, Aldershot 1991.
2. eg. W.Fulkerson et al., *Energy Technology R&D: What Could Make a Difference?*, Oak Ridge National Laboratory, ORNL-6541, Tennessee, US, 1990, Vols 1-3; UK Department of Energy, *Background Papers Relevant to the 1986 Appraisal of UK Energy Research, Development and Demonstration*, ETSU-R-43, HMSO, London, 1987; OECD/IEA, *Energy Technologies for Reducing Emissions of Greenhouse Gases*, Proc. Experts Seminar, Paris, 1989 (Vols 1 & 2); T.B.Johansson, B.Bodlund, R.H.Williams (eds), *Electricity: Efficient End-Use and New Generation Technologies, and Their Planning Implications*, Lund University Press, Lund, Sweden, 1989.

developments in technologies not covered in the case studies in this volume.

3.1 End-use technologies

Technologies for improving the efficiency of energy use or changing the fuels used span a diverse range. In transport, better engines, transmission, and vehicle and rolling characteristics can improve vehicle efficiencies, and this forms the first of the demand-side case studies, in Chapter 4. Alternative fuels are also possible, including methanol, ethanol, natural gas, electricity and hydrogen. Developments in batteries and gaseous storage systems, and in small-scale fuel cells for converting gaseous fuels into electricity for powering efficient electric vehicle drives, may improve the outlook for new vehicle fuels.

Draught-proofing and roof and cavity wall *insulation* are simple ways of reducing heat losses (as is education to encourage better occupant habits), still far from fully exploited. More advanced technologies for reducing heating requirements in buildings and some other applications include better *controls, vacuum and other special insulating panels,* and *insulating windows* including double panes with vacuum/gel spacers, and selective films which help to trap heat inside.

Heat can be supplied more efficiently by using *condensing boilers*, which exploit the latent heat in the steam from gas combustion; and *heat pumps*, which use refrigeration-type cycles to transfer heat from cooler external sources to warmer indoors. Also there are important opportunities for exploiting spare heat from a variety of sources, including power stations *combined heat and power* (CHP-district heating, also known as 'cogeneration') and localized small-scale generators (small-scale CHP).

Advances in *electric motor drives* can improve efficiencies in a wide variety of electric motor applications, which may be amplified by better optimization of systems especially for pumping. The savings may be particularly large when coupled with advances in *power electronics*, as applied to adjustable speed drives for applications with varying loads such as many pumping uses. The combination of better motors, insulation and design can more than halve consumption in many electrical appliances, with some of the most notable gains available in *efficient refrigeration,* which forms the second demand-side study, in Chapter 5.

New and improved lighting technologies include *compact fluorescent* lamps in place of traditional filament lamps, and *high frequency* operation of linear fluorescent lamps. Combined with better design and control, these can achieve still larger gains than the relative savings available in appliances. Lighting forms the third demand-side study, in Chapter 6.

A wide range of improved industrial processes is available. Various chemical and brewing industries separate different fluids by distillation, which is energy-intensive; *selective membranes* have now been developed which can separate fluids with far less energy use. Electrothermal techniques use focused microwave radiation to deposit intense heat at particular locations on surfaces or within materials, for example with welding and finishing, for relatively little energy input; laser techniques can also replace some energy-intensive processes, for example, for etching surfaces. New *catalysts* are continually being developed within the chemicals industry to accelerate reactions, and/or enable them to proceed at lower temperatures, again saving energy. *New materials* can also yield savings in both production and applications, as for example when strong plastics can replace metal, or smoother materials are applied to reduce friction. Many such options are reviewed by Goldemberg et al,[3] who also summarize new energy-saving processes for particular industries such as metals, pulp and paper. *Recycling* can also reduce energy requirements especially in the metals industries. The diversity of industrial activities makes potential improvement difficult to assess, but most industrial processes are as yet far from theoretical efficiency limits.

Broad applications of modern electronics in *sensors and control systems* for industry, *building energy management,* and *metering technologies* can optimize energy use; building energy management, particularly for service sector buildings, forms the fourth and final demand-side case study, in chapter 7.

3.2 Fossil fuel production and conversion

Techniques for extracting fossil fuels are by no means static. *Three-dimensional seismic mapping* enhances capabilities for searching and identifying fossil deposits, and reduces the risks of dry wells. A

3. J.Goldemberg, T.B.Johansson, A.K.Reddy, R.H.Williams, *Energy for a Sustainable World,* Wiley/Eastern, Delhi 1988, pp.132-48.

variety of techniques for *enhanced oil recovery,* raising the percentage of oil extracted from reservoirs above the typical levels of 20-35%, are available and some have been brought into commercial use even at existing prices.[4] Many more will be viable when oil prices rise again, helping to extend exploitable reserves. Continued development of *coal mining technology* has led to large reductions in employment and costs, a trend extended further by increased automation of underground operations, as well as by automated strip mining. Developments in techniques for *piping and liquefaction of gas* continue, and will contribute to the expansion of this fuel.

Various avenues for transforming the characteristics of fossil fuels, beyond traditional coal washing and oil refining, are also in use or under investigation. Processes for *liquefying coal* have been used at times and in places of severe oil shortage; attempts to improve these techniques formed one of the major R&D responses to the oil shocks, though coal liquefaction is not now expected to be competitive with oil until oil prices exceed about $40/bbl. Deposits of very heavy oil in Venezuela are however being converted into the commercial product of *orimulsion* as a liquid competitor to coal. Liquid fuels are themselves transformed into ever more specialized products, ranging from traditional petrochemicals to *reformulated gasoline*, a product with altered chemical composition which is now being marketed in the US to provide cleaner-burning car fuels. *Gas synthesis* techniques for producing methanol, ethanol and other liquid fuels from natural gas are also still advancing, with increasing conversion efficiencies and falling costs.

For producing electricity from fossil fuels, the steam turbine powered by oil or pulverized coal has reached its zenith, and a host of other candidates have emerged to challenge its dominance. The efficiency and performance of industrial *gas turbines* have improved dramatically, and gas turbines derived from aero-engines are also being applied and improved for electricity generation. By linking gas turbines with steam turbines in *combined cycles,* the efficiency can be raised above that of either independent component, to unprecedented levels. Similar performance can be achieved by other modifications of gas turbine cycles. Gas turbine systems are taken as the first of the supply-side case studies, in Chapter 8.

---

4. For detailed discussion see P.Simandoux, D.Champlon and E.Valentin, *Status and Outlook of Improved Oil Recovery*, Institut Français du Pétrôle, 1989.

These technologies are not confined to natural gas: systems for *coal and biomass gasification* for driving gas turbines are being developed. This offers one route (demonstrated, but not yet commercially viable) to cleaner and more efficient coal-based generation, but there are many others, including various forms of *fluidized bed* combustors, and *fuel cells*. Clean coal technologies are examined in Chapter 9.

3.3 Nuclear technologies

Research and development of improved nuclear technologies continues. The dominant light water reactor has many variants, and designs are still evolving. Other existing designs include the Canadian CANDU reactor. Other approaches are being studied in response to the industry's difficulties, including much smaller-scale, *passively safe reactors* such as the modular high temperature gas reactor. In these systems the core is small and/or diffuse enough to be cooled by natural convection if the cooling system fails. As yet however such systems exist only in design and commercial prototypes are not being developed.

Research programmes on the *fast breeder reactor*, which is fuelled by plutonium and able to breed plutonium fuel from natural uranium, continue albeit at a much reduced pace. The only full-scale FBR developed, the French Superphoenix, has had a troubled history. The UK and US programmes are being wound down in favour of broad international collaboration, which may result in attempts to construct a more commercially viable prototype, but it is not expected to be a commercial proposition before the second quarter of the next century.

Fusion power, in which light atoms are fused to release energy (rather than heavy atoms split as in fission), continues to attract headlines. Hope for *cold fusion* dissipated over the year following the initial press announcement in March 1989, as other laboratories proved unable to reproduce the claimed results of Pons and Fleischman.[5] *Magnetic confinement*, which attracts the bulk of international fusion R&D funding, received a boost with the long-awaited announcement that the JET test reactor had produced significant power from the reaction.[6] However, essential problems of economic and environmental feasibility

5. Frank Close, *Too Hot to Handle: The Race for Cold Fusion*, Allen/Princeton University Press, 1990.
6. 'Fusion becomes a hot bet for the future', *New Scientist*, 16 Nov. 1991.

remain, and the conclusions of the most in-depth official review[7] - that fusion cannot plausibly become commercially feasible much before the middle of the 21st Century and that even this is highly uncertain - remain unaltered.

3.4 Geothermal and renewable energy technologies

Useful heat can be extracted from the earth's crust in the form of geothermal energy. *Hot aquifers* are the simplest to exploit, and are already in widespread use where hot waters are near the surface, with several thousand megawatts installed worldwide. Other more speculative options include the exploitation of deeper *geopressurized brine* and the extraction of heat from *hot dry rocks* and *magmas*. In 1991 the UK government decided that the prospects for hot dry rock (HDR) were not sufficiently good to justify extending the UK programme to the depths which would be required for significant power generation in the UK. The US and the European Community also have HDR programmes.

Hydro power has traditionally been exploited through large dams, but some can also be tapped through much smaller *micro-hydro* schemes using small dams or run-of-river techniques.

Technical advances have served to 'modernize' a wide range of other renewable energy technologies.[8] Technologies for extracting *wind energy* have developed rapidly and are now the closest to widespread commercial realization of all the modernized renewable technologies; wind energy is taken as the third supply-side case study, in Chapter 10.

Direct solar energy provides the greatest of the renewable resources. It can be converted to electricity by various technologies. *Solar power towers* use mirrors to focus sunlight on a central tower to heat a thermal turbine system. Various demonstration schemes have highlighted potential difficulties associated with the thermal stresses and thermal inertia involved, though for desert regions they may be promising.

---

7. Office of Technology Assessment, *Starpower - the US and the International Quest for Fusion Energy*, OTA, Washington, 1987.
8. An extensive recent review and analysis of renewable energy technologies and their prospects in the US is given in US DoE, *The Potential of Renewable Energy - an Interlaboratory White Paper*, SERI/TP-260-3674, 1990. Most promising in-depth studies will be presented in R.H.Williams, T.B.Johansson, A.K.Reddy (eds), *Fuels and Electricity from Renewable Sources of Energy*, Island Press, Washington DC, forthcoming 1992.

Distributed thermal systems use curved mirrors to reflect sunlight on to tubes which carry working fluids that drive a turbine. These smaller and more modular systems have been deployed commercially in southern California, sometimes in combination with gas supplies to drive the turbine when insufficient sun is available, and continuing improvements are predicted. However the technology with the broadest potential application, because it does not rely on direct sunlight and can be used at almost any scale, is *photovoltaic solar cells*. These are already used in many special applications and are developing rapidly, and photovoltaics is taken as the final case study, in Chapter 11.

Solar energy can also be used directly for *solar heating,* by using better building design and in greenhouses, and for drying, for example agricultural crops. Sunlight already contributes to energy supplies in these ways, and options for increasing that contribution are frequently relatively simple and inexpensive. At the opposite end of the development scale, a range of techniques for direct chemical processes to convert sunlight into fuels is being explored.

The natural equivalent to synthetic chemical conversion is biomass energy. Biomass is available in a wide variety of forms from urban waste and crop residues to forest plantations. Traditionally such biomass is dumped, burnt in open fires, or rarely, used to raise steam in boilers for electricity. *Biomass gasification* for driving gas turbines is being explored as a much more efficient approach, and *enzymatic conversion* can improve ways of converting biomass to liquid fuels such as ethanol. These two routes can be up to ten times as efficient as open-hearth burning. Complete operating systems have not yet been demonstrated, but the prospects are promising especially for cheap biomass wastes, or if used in connection with techniques being explored for increasing yields from 'energy crops'.

Tidal motions can be converted to power through *tidal barrages*; a 240MW scheme has been operating in France for over 25 years and other smaller schemes exist, but the long timescales and capital outlay required, together with uncertainty about the environmental impacts, have deterred exploitation of potentially very big schemes for example in Canada and the UK. There have also been proposals for exploiting deeper tidal streams and other ocean currents. Devices for tapping *wave energy* at the shore have been demonstrated and sold; a wide variety of options for floating systems to exploit the much greater resource in deeper waters have been proposed and some demonstrated in small-scale prototypes, but with no full-scale demonstrations, uncertainties about the costs and viability of offshore wave energy persist.

Ocean thermal energy conversion systems, to exploit the temperature differences in tropical surface waters have been developed, but appear costly and there are concerns about the marine impacts. Other more speculative renewable energy options include *space-based solar power,* the exploitation of energy from *vapour pressure differences* in deserts (dew point energy), and *salt gradients* where fresh water meets the sea.

3.5 Criteria for case studies

In short, an enormous range of options for new or improved technologies for energy production and use have been proposed and studied, with many demonstrated in some form. They are at varying stages of development. Examining each in depth is not possible within the confines of one book or project, and is not necessary for drawing out key issues concerning the potential impact and implications of emerging technologies. In this study therefore we focus upon selected case studies.

Various criteria were used for selecting case studies. First, they are restricted to technologies which are already beginning to make an impact on energy markets, or which could start to do so during the 1990s. By the standards of most technological developments in energy production and use, this is a short focus. It restricts the scope to technologies which are already in a fairly advanced stage of development, and which are at least beginning to knock on the doors of the energy market place. Thus the study does not consider 'blue skies' options such as nuclear fusion, or hundreds of other more modest technologies which have been suggested as part of long-term solutions to energy problems, but for which the real costs and characteristics remain highly uncertain. This relatively short focus means that many features of the technologies are clear, and that the unavoidable technological uncertainties of looking ahead need not swamp the analysis. The studies themselves are not all restricted to the 1990s, but extend the analysis to consider the possible longer-term role of the technologies.

Second, the technologies are all of potentially broad enough application to have a significant impact on energy demand or supply, if not actually during the 1990s then in the decade or two after. Thus whilst many minor process improvements may be important for particular energy-using industries, and ways of enhancing the energy extracted from particular mines or reservoirs may extend the production lifetime, the focus of this study is on technologies which could have broader implications for national energy balances and the competitive position of different sources.

Third, studies are chosen which reveal significant lessons for public and industrial policy.

Even within these confines the project does not seek to examine all possible candidates. Rather than attempt to explore the full range of options, we have sought to focus on particular examples which illustrate key points, and to examine them in some depth. Thus this book focuses upon eight case studies of technologies which are emerging or which could emerge into energy markets during the 1990s.

The growing importance of understanding the demand side of the energy balance is reflected in the equal division of case studies between demand and supply technologies. *Vehicle technologies* are chosen as the dominant technology in what is the fastest-growing of all areas of energy demand, and the one in which oil is central. It also illustrates interactions with non-CO_2 environmental factors, and important issues of both real and imagined trade-offs against various non-economic considerations, together with the complexities of trying to stimulate efficiency improvements in the face of them.

Efficient domestic appliances addresses the peculiarities of a major demand sector in which there are apparently efficiency gains available from simple improvements with few if any identifiable trade-offs or particular sensitivities involved on the part of the consumers involved. It illustrates the scale of improvements available and some of the means which might be used to tap this potential.

Efficient lighting examines a situation in which very large efficiency gains are available by a move to more sophisticated technology, but in which there are also non-economic trade-offs arising from the very different characteristics of some of the new lighting technologies, and a barrier of much higher first costs for some (though not all) of the changes involved.

The final demand side study examines overall *building energy management*; this highlights the importance of addressing the whole system that is using energy and the potential of modern technologies further to integrate and optimize energy use from many different components, and illustrates the complexities of focusing managerial attention in small and very diverse businesses on the opportunities available for low-cost energy savings.

Industrial energy demand is not addressed in this book for various reasons. Gaining detailed and accurate information about advanced industrial processes which might save energy can be difficult, in part because of commercial sensitivities. The diversity of the industrial sector means that experience captured by examining one particular process might not validly extend to other industries. Also, for the heavy industrial processes which contribute most to energy demand, energy cost is already a potent indicator of which managers are already well aware, so that the scope for improvements driven by policy changes (other than pricing) may be very limited.

On the supply side, a study of *gas turbines* examines the technologies which are increasingly being billed as the foundation of future electricity expansion in many areas of the world: why they are attracting such excitement, how they came to their present position, the impact they may have, and the extent to which this will depend on national and regional circumstances.

In many areas of the world however, coal remains the main domestic resource and in many countries it will remain a crucial fuel, but it is also the dirtiest one. The study of *clean coal technologies* reflects the cumulative impact of growing environmental concerns and competition from other fuels on one of the biggest and most conservative areas of heavy engineering, namely the generation of electricity from coal, and considers the implications of the major technical advances being made.

Of all the modernized renewable energy technologies, *wind energy* has probably shown the most spectacular advances, and is the closest to supporting a viable commercial industry competing for the main grid-connected supplies. The history of wind energy reveals crucial lessons about the importance of government policy and utility organization, for good and for bad, in nurturing emerging renewable technologies. It illustrates the surprises and problems engendered by a technology which differs so radically from conventional power generation in scale, industrial sponsorship, and output characteristics.

Finally, a study of *photovoltaic solar cells* examines the technology which many consider the renewable technology with the greatest long-term potential as the foundation of sustainable energy supplies; a technology which paradoxically mixes the extremes of high technology production processes with great simplicity in application, and large economies of scale in production with very small unit sizes in many diverse applications.

As indicated above, this selection is not intended to be comprehensive. However, the selection does include several of those which seem likely to have the greatest impact on future energy markets, and which may have the most to offer in responding to the growing pressures on energy systems, most notably environmental pressures. The final part of the book then draws together the case studies and explores their broader implications for future energy developments and the role of government and international policies.

PART II: DEMAND-SIDE TECHNOLOGIES

Clean and Efficient Cars

Energy consumption for transport is currently growing faster than in any other sector of the global economy, and vehicles are a major source of pollution. This chapter examines the potential for advanced technologies to reduce emissions and energy consumption. Many of the technologies discussed are applicable to all types of road vehicles, but the analysis focuses on cars as the dominant issue.

A number of factors influence how much energy is used by vehicles including: engine size and efficiency, vehicle design, drivers' skill, vehicle speed, level of maintenance, distance travelled and road conditions. Reducing energy demand in the sector as a whole will require a combination of measures including traffic restraint, promotion of public transport, walking and cycling, as well as an introduction of new vehicle technology.

Despite steady improvements, especially during the time of high oil prices, many options remain. These include: greater use of diesel engines; modified engine design and controls; transmission improvements including continuously variable transmission; and reductions in weight, and in rolling and air resistance. Available technology could deliver up to 40% improvements in average fuel economy for new cars by the year 2010 at little extra cost, without loss in performance, and despite the introduction of more stringent emissions controls at the same time. However, manufacturers are estimating much smaller increases.

A number of market barriers exist to the introduction of fuel efficient vehicle technology, particularly motorists' lack of interest in fuel economy at current fuel prices and their apparently insatiable desire for increased car performance. It is unlikely that market forces alone will lead to the introduction of energy efficient technology despite its cost-effectiveness. Government intervention is required to stimulate the market both in supply and demand. Measures could comprise regulatory standards and incentives for encouraging the industry to develop new energy efficient technology, and fiscal incentives (eg. tax incentives for energy efficient vehicles) to encourage motorists to buy fuel efficient, economical vehicles. Special attention may be required concerning the transfer of old and inefficient vehicles to developing countries.

In OECD countries energy use for transport already accounts for nearly a third of energy consumption, and is growing faster than for any other end-use sector. Road transport accounts for over 80% of the energy used by the transport sector as a whole.[1] In 1991 there were over 550 million vehicles in use around the world, of which over 400 million are cars. If historic rates of growth are maintained, the global vehicle population could exceed one billion within the next 20 to 40 years.[2]

Motor vehicles are a major source of air pollution, probably larger than any other single human source. They are responsible for much of the deterioration in urban air quality seen in cities around the world in recent years, and for photochemical smog, which can stretch over large areas in hot sunny weather. Road vehicles also make a significant contribution to acid rain and global warming.

1. G.McInnes, 'The Effectiveness of Policy Measures to Promote Energy Efficiency in Road Transport', paper presented at Tomorrow's Clean and Fuel-Efficient Automobile: Opportunities for East-West Cooperation, an international conference organized by OECD, IEA and ECMT, Berlin, March 1991.
2. The slower growth rate is based on an extrapolation of historic trends in global motor vehicle registrations since 1970; the faster figure is based on historical trends in registrations per capita since 1970, and is considered an upper limit. James J.Mackenzie and Michael P.Walsh, *Driving Forces: Motor Vehicle Trends and Their Implications for Global Warming, Energy Strategies, and Transportation Planning*, World Resources Institute, December 1990, Washington DC.

Although new cars are more energy efficient now than they were in the 1970s, energy consumption from road transport has been growing as a result of an increasing number of vehicles being driven further. This growth in traffic has in the past outstripped new technological developments. Current forecasts for traffic growth suggest that this will continue and energy consumption in the sector will continue to grow unless there is some form of traffic restraint, particularly in the OECD countries.

High levels of traffic lead to congestion which also increases energy consumption, as vehicles are driven inefficiently during the stop-start conditions that characterize congested roads. It is now becoming widely accepted that building new roads does not reduce congestion but rather releases suppressed demand and thus traffic restraint measures, such as road pricing, increased fuel prices, area bans and severe parking restrictions are increasingly being considered as solutions. However, for traffic restraint to be socially and politically acceptable, people must be offered an alternative mode of transport. Thus improvements in facilities for pedestrians, cyclists and public transport passengers/goods must be provided. In many of these areas there are a number of emerging technologies that could help reduce the energy demand of the transport sector as a whole. These include the use of 'smart card' technology for road pricing, 'guided' routes for buses, and electronic information systems for buses and light railways.

There are a number of other factors that influence how much energy is used by vehicles including the drivers' skill, and the level of maintenance. New technologies have a role to play in these areas as well, for example electronic cameras can make the detection of vehicles travelling at excessive speeds easier, while on-board diagnostics can inform motorists of faulty spark plugs or when the tyres need re-inflating.

Thus a multi-faceted approach is needed to tackle the road transport problem, and new technology has a role to play in all areas. However, this chapter focuses on emerging energy technologies for road vehicles. Cars dominate road traffic, accounting for three out of every four motor vehicles on the road, and they consume a similar proportion of road transportation fuel. This is unlikely to change in the foreseeable future, especially in the OECD countries where the motor car is, by far, the major mode of passenger transport. Even a radical policy to encourage modal shift to more energy efficient public transport is unlikely to have a large effect on energy consumption - even a doubling of patronage, which would have major implications for the public transport infrastructure, will make only a relatively small impact on the total passenger car mileage.

The potential for improving the fuel consumption of new cars is thus a central issue, on which most of this chapter concentrates, focusing upon those technologies that are already in some production vehicles, or are close to being introduced into the market. However, the clean-up of non-CO_2 emissions has an important impact on car design. These issues are discussed first.

4.1 Cleaning up emissions: end-of-pipe and alternative fuels

Pollution from cars is not a new problem. Measures to improve combustion have been undertaken for decades, helping to reduce smoke and gaseous emissions, but some emissions from the engine cannot be avoided. In the 1970s, concerns in the US and Japan grew sufficiently strong that these countries introduced catalytic converters to remove pollutants from exhausts.

Catalytic convertors increase vehicle costs, and there has been considerable debate over the potential energy penalty of using them, with estimates ranging up to about 10%.[3] However, in introducing catalytic convertors, other changes have to be made to the vehicle including more advanced engine ignition. These can offset the losses, and some European motor manufacturers anticipate no fuel penalty or even a slight improvement.[4]

Even the adoption of three-way catalytic converters however has not proved sufficient to allay environmental problems in the face of continued traffic growth. In the US, where yet still more stringent emissions controls have recently been agreed there may be other energy penalties, depending on the technology used to meet the new limits. In southern California, oil companies in 1989 launched 'reformulated gasoline' which has a higher oxygen content to make it burn more cleanly

3. See for example, D.J.Martin and R.A.W.Shock, who suggest that the energy penalty lies between 2% and 9%. *Energy Use and Energy Efficiency in UK Transport up to the Year 2010*, Energy Technology Support Unit, Department of Energy, HMSO, 1989.
4. Volkswagen Audi claimed in 1990 'Audi and VW experience has shown that minor increases in some parts of the performance spectrum are matched by decreases at others. The overall effect is negligible' ('Catalytic Converters and Unleaded Fuel', a briefing note, 6th edition, Volkswagen Audi Press Office, 1990).

and other characteristics such as less evaporation and reduced aromatic compounds. However, the improvements do add significantly to the price of the fuel.[5]

One frequently mentioned means of reducing emissions from road traffic is switching to alternative, non-petroleum fuels such as natural gas, alcohol fuels (ethanol and methanol), electricity, and hydrogen. This option is attractive in that it also reduces transportation's dependence on petroleum fuels. However, there are a number of technical problems yet to be resolved with these new fuels, and their environmental benefits are not as clear cut as their proponents would suggest.

Electric vehicles have been widely promoted in the media since the late 1980s, but further major developments in battery technology are required before there is any prospect of them approaching the performance of petrol vehicles. If motorists are willing to accept the poorer performance of current electric vehicle technology, then an alternative approach may be to develop a petrol or diesel vehicle with similar performance. This hypothetical vehicle would require a very small engine and fuel tank, making it light, and extremely fuel efficient.

Ethanol, usually derived from biomass, is already mixed with petrol in some areas of the world, and in Brazil many vehicles run on ethanol alone. However, ethanol remains expensive, and in many areas biomass production on the scale required to displace all petrol use is not a viable option, although on a local scale it could be a useful fuel especially when mixed with petrol. However, it is important to ensure that the additional pressure of fuel crops on available land does not result in the use of marginal lands and intensive agricultural practices to the detriment of the environment.

It is unlikely that alternative fuels will have a major impact on energy consumption over the next decade (during the timescale under discussion in this chapter). During this period alternative fuels are likely to be restricted to areas with very poor air quality such as south California, where regulations have mandated a market share for 'zero emission' vehicles rising to 10% of new car sales by 2003. On a longer timescale, the use of electricity generated from renewable energy sources is an environmentally attractive option both for electric vehicles and for the

5. The 'Federal Formula' for reformulated gasoline adds 5-15% to the *costs*, whilst the tougher 'Californian' formula adds 15-30% to the costs (Arnold Baker, ARCO, private communication).

production of hydrogen fuel. There may also be some potential for vehicles to be powered by electricity produced on-board in fuel cells, probably from methanol or hydrogen. Another possibility in some climates is solar-photovoltaic vehicles with battery storage.

Emissions of both the traditional pollutants associated with energy production and carbon dioxide from the use of these new fuels depend to a large extent on their feedstocks and methods of production. An analysis from the US suggests that doubling the vehicle fleet fuel efficiency would do more to reduce CO_2 emissions than a switch to all alternative fuels except the more speculative and distant options drawing on non-fossil electricity and biomass.[6] Improved efficiency, because it would also reduce the overall fuel burnt, could also contribute significantly to reducing some other pollutants (although nitrogen oxide emissions might be increased).

4.2 Trends in vehicle efficiency

Since the first oil price shock of 1973 the fuel efficiency of road vehicles has improved markedly. Most of the improvements occurred between 1975 and the early 1980s when oil prices were still rising and there was strong consumer interest in saving energy. During this period motor manufacturers were forced by legislation and voluntary agreements to improve fuel consumption. In the USA they did this by reducing vehicle weights, while in Europe they introduced new technology to improve fuel economy because this is what the market demanded. In the decade from 1978 new car fuel economy improved by about 20% in the UK[7] and similar improvement was made in other European countries. In the US fuel economy improvement of 36% was reported over this period albeit from a much lower starting point.[8]

---

6. M.A.DeLuchi, D.Sperling and R.A.Johnston, *A Comparative Analysis of Future Transportation Fuels*, Institute of Transportation Studies, University of California, Berkeley, California, 1987.
7. National Audit Office, *National Energy Efficiency*, HMSO, 1989.
8. R.M.Heavenrich and J.D.Murrell, *Light Duty Automotive Technology and Fuel Economy Trends Through 1989*, US Environmental Protection Agency, Washington 1989.

The average fuel consumption of new cars in the US today is 8.4 litres/100km, while in other OECD countries it is slightly lower, for example, in the UK it is 7.4 litres/100km.[9] It is lowest in Italy (6.8 litres/100km), where there is a predominance of small cars, many of which are diesel fuelled. In many countries the fuel consumption of new car fleets has actually been rising in the last few years due to increased demand for larger, higher performance cars. Data on the fuel economy of new cars in developing countries are poor, but it is probably somewhat lower than that of OECD countries.

These improvements in new car fuel efficiency have not generally been reflected in on-road fuel consumption. In Europe and Japan the fuel saving has been around 10% on a vehicle kilometre basis and, due to declining load factors, less on a person kilometre basis. On-road fuel consumption trends to 1988 are illustrated in Figure 4.1.[10]

In the developed countries there is a relatively rapid turnover of the vehicle stock, with about 10% being renewed every year. In the developing countries, on the other hand, vehicles are typically much older, and there is a substantial market for imported second-hand cars. Thus in these countries it will take longer for innovative technology to penetrate the market, and the on-road reduction in fuel consumption is likely to be even slower.

4.3 Technical options for improving efficiency

There are basically two approaches to reducing the fuel consumption of a vehicle. Firstly, the power required for propulsion must be produced most efficiently; and secondly, the power required to propel the vehicle must be minimized. Out of a world fleet average fuel consumption of about 10 litres/100km, it has been estimated that about 75% of the fuel's

9. Fuel consumption in this chapter is presented in litres per 100 kilometres (litres/100km). A vehicle with a fuel economy of X mpg (imperial) has a fuel consumption of (282.5/X) litres per 100 kilometres; while one with a fuel economy of Y mpg (US) has a fuel consumption of (235.3/Y) litres per 100 kilometres. Note that fuel economy is the reciprocal of fuel consumption. Fuel economy is measured on different test cycles in Europe, USA and Japan. For example, European fuel economy figures include a higher speed component, and therefore these figures are not directly comparable. These official fuel economy figures under-estimate actual on-road fuel consumption.
10. Lee Schipper, 'Improved Energy Efficiency in the Industrialised Countries, Past Achievements, CO_2 Emission Prospects', *Energy Policy*, March 1991, p.128.

Figure 4.1 Fuel consumption trends to 1988

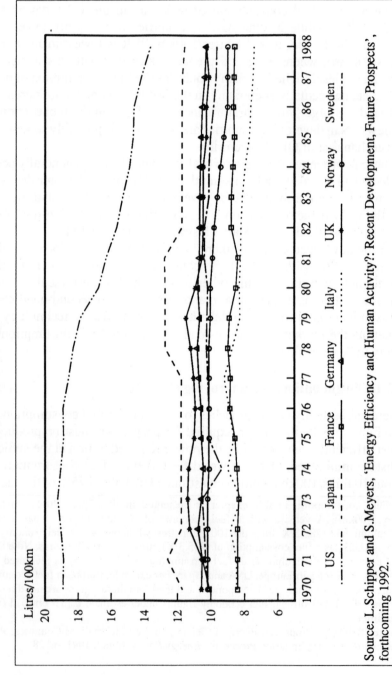

Source: L.Schipper and S.Meyers, 'Energy Efficiency and Human Activity?: Recent Development, Future Prospects', forthcoming 1992.

Table 4.1 Technologies available for improved car fuel economy

Technology design approach	Potential gain (as % of 1986 petrol car)
Engine improvements	
Improvements to current petrol engines (eg. precision cooling, reduced engine friction, reduced pumping losses)	up to 6
4 valve per cylinder 4 stroke lean burn petrol engines	5 to 15
Direct injection 2 stroke petrol	0 to 10
Diesel engines (conventional and improved direct injection high speed designs)	20 to 30
Electronic engine management	5 to 20
Vehicle and transmission design improvements	
Transmission improvements (eg. automated manual)	10 to 15
Continuously variable transmission	10 to 15
Weight reduction	15 to 20
Aerodynamic improvements	5 to 10
Improved tyres, lubricants, accessories	5 to 10

Source: D.J.Martin and R.A.W.Shock, *Energy Use and Energy Efficiency in UK Transport up to the Year 2010*, Energy Technology Support Unit, Department of Energy, HMSO, London, 1989.

energy is lost in the engine, 15% in the transmission and elsewhere, and only 10% turns the wheels.[11]

Table 4.1 lists the potential fuel economy benefits of some of the available technologies relative to a 1986 petrol engined car. The estimates given are for each individual technological improvement alone. A combination of the measures is not necessarily additive, and some of the developments listed are mutually exclusive.

11. R.W.Melde, I.M.Maasing and T.B.Johansson, 'Advanced Automobile Engines for Fuel Economy, Low Emissions and Multifuel Capability', *Annual Review of Energy*, Vol.14, 1989, pp.435-44.

Engine improvements

Diesel engines are more efficient than petrol engines for a number of reasons.[12] Firstly, and most importantly, they operate at a higher compression ratio. Secondly, they use leaner air-fuel mixtures, that is, the air-fuel ratio is lower. Thirdly, the engine output is controlled by metering the fuel supplied to the fuel injector, thus avoiding the part-load pumping losses of petrol engines. In addition, diesel engines are generally designed to operate at lower speeds, and consequently the frictional losses are lower.

Indirect injection diesel engines can offer fuel economy improvements of up to around 30% compared with traditional petrol engines,[13] while direct injection diesel engines may in turn use 10-15% less fuel than indirect injection models.[14] However the major disadvantage of diesel engines is that they have a lower specific power output and thus diesel vehicles have poorer performance compared with a similar petrol model, although the use of power boosters can overcome this drawback. Other disadvantages include higher capital cost, worse noise and vibration and higher emissions of nitrogen oxides compared to a petrol engine with a three-way catalyst.

Petrol engines can operate more efficiently by using power boosters such as turbochargers to enable reduced engine size, rather than to increase power output; electronic engine management to ensure that the vehicle is running as efficiently as possible; and reducing engine warm-up time. The latter is important as fuel consumption increases substantially when a vehicle is cold.

---

12. It should be noted that while diesel engines are more efficient than petrol engines, the former is a denser fuel and the benefits in terms of reduced carbon dioxide emissions is less than would appear from a straight comparison of fuel consumption unless the whole fuel cycle emissions are taken into account. More energy is required to refine petrol than diesel, and when this is taken into account, it is generally assumed that comparison of the fuel consumption does in fact approximately equate to a comparison of carbon dioxide emissions and primary energy consumption.

13. J.M.Dunne, 'A Comparison of Various Emission Control Technology Cars and Their Influence on Exhaust Emissions and Fuel Economy', *Warren Spring Laboratory Report No LR 770 (AP)*, Warren Spring Laboratory, Stevenage, UK.

14. C.A.Amann, 'The Automobile Engine - A Future Perspective', Paper presented to the Future Transportation Technology Conference and Exhibition Society of Automotive Engineers , Vancouver, 8-10th August 1989.

Considerable effort has also gone into improving compression ratios and lean burn combustion technology. The improvement of lean burn combustion technology enables air-fuel ratios to be reduced with a consequent increase in efficiency. However, emissions of nitrogen oxides are higher than from a conventional petrol engine fitted with a three-way catalyst, and cars with lean burn engines are unable to meet the forthcoming European emissions legislation.[15] The future of this technology is dependent on the development of a catalyst capable of removing NO_x from the oxygen-rich exhaust. A number of car manufacturers are currently investigating the technology.

Vehicle and transmission design improvements

Energy efficiency can also be improved by reducing vehicle weight. Substantial progress has already been made in this area, but vehicle weight could be reduced further by material substitution through, for example, the extensive use of plastic materials for structural, body and engine components and the wider adoption of aluminium engine blocks.

Substantial improvements in fuel efficiency are also possible by using transmissions which enable the engine speed to be matched more closely to the vehicle load. Using more flexible transmissions enables the engine to operate at a speed allowing maximum efficiency under a particular set of vehicle conditions. One such technology already used in some production vehicles is continuously variable transmission. As the name suggests the discrete steps associated with conventional gearbox systems are eliminated and the range of gearing ratios is also much wider.

Overcoming the vehicle's resistance to aerodynamic drag, especially at higher speeds, is yet another way of improving fuel efficiency. At present the best European cars have a coefficient of drag of 0.28; but at least one manufacturer believes that its company's average will be down to around 0.25 by the turn of the century, giving a 10% saving.[16]

---

15. The European Community's consolidated Directive (91/441/EEC) introduces mandatory emissions limits for cars for 1992. The limit values were set at a level that can currently only be met using a controlled three-way catalytic convertor. These limit values are broadly equivalent, but possibly slightly more stringent than the US Federal standards. California has its own more stringent standards. The US and Europe use different test cycles for measuring emissions and thus direct comparison is difficult.
16. D.L.Bleviss, *The New Oil Crisis and Fuel Economy Technologies: Preparing the Light Transportation Industry for the 1990s,* Quorum Books, New York, 1988.

4.4 Prototype vehicles

In the mid-1980s a number of low-energy concept vehicles were developed. These prototypes generally used a direct injection diesel engine and incorporated many of the technologies discussed above, but few were developed with production in mind and none to date have been produced commercially. However, some of the technologies have been incorporated into current models such as the Citroen AX range which utilizes much of the advanced technology developed under the Peugeot low energy prototype development programme.

Some of these low energy prototypes have a fuel consumption of less than 2.5 litres/100km when driven a steady 90km per hour[17] compared to the best production vehicles of 3.6 litres/100km. This demonstrates that it is feasible to produce cars with significantly better fuel efficiency than can be purchased today, although they generally do not have the level of comfort, safety, and emissions control that would be required for a production vehicle. These prototypes do not even represent the bounds of technical feasibility since none utilize all the energy efficiency technologies that have been developed in recent years. Some idea of the scope for further progress is suggested by the world record of 6,409 miles on a single gallon of petrol, albeit using essentially a converted bicycle with a 50cc engine.[18]

Of particular interest amongst the prototype vehicles are Volvo's LCP (Light Component Project) 2000 experimental cars as these vehicles were built to meet current and projected emissions and safety standards. The experimental cars have a total energy consumption which is less than 60% of that of similar sized conventional cars, with a fuel consumption of 3.6 litres/100km in mixed driving. It is predicted that production costs will be no higher than for other comparable sized cars.[19]

17. Official fuel consumption figures are not generally repeatable in real driving conditions. For information on low energy prototype cars see Bleviss, ibid.
18. Equivalent to 0.04 litres/100km. D.R.Blackmore and G.B.Toft, 'Shell's Mileage Marathon Competition', Paper presented to the Institution of Mechanical Engineers Conference on Automotive Power Systems - Environment and Conservation, Chester, 10-12 September 1990. The author is not suggesting that this type of vehicle would be either acceptable or desirable as a replacement of current cars, but it suggests that by redefining the concept of passengers cars a fuel consumption of, say, 1 to 2 litres/100km may be achievable.
19. R.Melde, 'Volvo LCP 2000 Light Component Project', *Paper No 850570*, Society of Automotive Engineers, Warren Dale, PA, 1986.

Table 4.2 Fuel efficient production cars on the UK market

Manufacturer	Model	Fuel	Composite fuel consumption, litres/100km
Daihatsu	Charade	Diesel	4.35
Citroën	AX14D	Diesel	4.41
Peugeot	205D	Diesel	4.63
Citroën	AX10	Petrol	4.71
Rover	Montego	Diesel	4.71
Renault	5TD	Diesel	4.79
Ford	Fiesta 1.8D	Diesel	4.87
Vauxhall	Nova 1.5D	Diesel	4.87

Note: the composite figure is based on the formula used by the Department of Transport, ie. the composite fuel consumption figure is made up of 40% urban test cycle, 50% 90kph (56mph) test and 10% 120kph (75mph) test.

Source: Department of Transport, *New Car Fuel Consumption: The Official Figures*, HMSO, October 1990.

4.5 Fuel efficient production vehicles

Cars capable of travelling an average of 4.4 litres/100km in official economy tests are currently on the market. Table 4.2 shows the best production cars on the UK market in 1990.

These figures represent a substantial improvement on average fuel consumption figures. For example, the average UK figure in the late 1980s was 7.4 litres/100km; the best average was for Italy at 6.8 litres/100km. The figure for the Citroen AX10 shows that a petrol engine can have very good fuel consumption, but the diesel AX14D, which is the diesel equivalent of the AX10, has a 9% lower fuel consumption.

4.6 Trade-offs: efficiency, emissions, performance, and costs

Motor manufactures have to balance a number of factors when designing cars. For example, engines can be optimized for emissions, performance or economy but not all three at the same time. Thus, it has been suggested, higher energy efficiency cannot be achieved whilst introducing more stringent emissions controls, and maintaining current levels of performance. In addition, it has been argued, improving energy efficiency through reducing vehicle weight will compromise vehicle safety.

There has been considerable debate in recent years over the energy penalties of different emissions control strategies, particularly in Europe, where the introduction of much more stringent emissions controls has been under discussion. The UK motor industry, which has made a large investment in the development of lean burn engines, have suggested that, for small cars, this technology offers the best compromise between energy efficiency and emissions reduction. Early development work suggested that these engines would give substantial energy savings, but in reality, the fuel savings have not been very high, with possibly a maximum gain of 5% but probably much less. Against this small energy efficiency gain must be balanced the higher emissions of nitrogen oxides from lean burn engines compared to those from conventional petrol engines fitted with catalytic emissions control.

Nitrogen oxides emissions are also a problem with the new highly energy efficient direct injection diesel engines that have been used in one or two production models in Europe. The energy gains to be made using these engines are much greater than those from lean burn technology. For example, a direct injection diesel engine can give a 2.0 litre family saloon car the fuel consumption of an economical small car. Given the huge potential for this technology, which has been proven in production vehicles, there may be grounds for allowing more time for new emissions control strategies to be developed before non-compliance with emissions limits forces these vehicles off the market.

Much of the new technology, originally developed to increase energy efficiency, has been used instead to improve vehicle performance. The average power output of new petrol engines is now about 50% higher than it was a decade ago. Much of this power is excess to normal driving requirements, and it has been estimated that, on average, a car works at only about 20% of its maximum power output.[20] Having so much excess power available makes fuel consumption very vulnerable to drivers' behaviour. For cars to retain comparable performance this improvement in power output could have facilitated a reduction of a least 20% in engine capacities; instead of which the average engine size has increased over that same period in the OECD countries.

During the 1970s, following from the oil crises, motor manufacturers, particularly in the USA, initially responded to the demand for energy efficient vehicles by reducing car weight. Not surprisingly, they became more susceptible in accidents, and reducing vehicle weight gained the reputation of compromising safety. However, in recent years, motor manufacturers have been looking increasingly at new materials and structural designs that can reduce weight while still maintaining vehicle strength and these could ensure that low energy cars are as safe as their heavier and more energy inefficient counterparts.

There may also be extra costs associated with the purchase of energy efficient technology. For example, diesel engined cars have traditionally been more expensive than their petrol counterparts, and financial savings have only been realized for high mileage motorists.[21] However, it is often difficult to obtain information on the costs of different vehicle technologies. In addition, some manufacturers may choose not to pass the whole cost onto the consumer, while others may increase prices well above the actual cost of installing the technology, as was the case with some car models when catalytic converters were first introduced into the UK.

Figure 4.2 shows a model-based calculation of the impact of improving efficiency, for a given base vehicle characteristics, on the *total* costs of owning and operating a vehicle in the US. This is a simplified study; it has been argued by, for example, Bleviss,[22] that making improvements to existing models in small incremental stages tends to be more expensive for the fuel savings achieved than redesigning the vehicle all together, and conditions differ substantially between countries. As mentioned earlier, it has been predicted that the Volvo LCP 2000 experimental cars could be produced at the same cost as comparable-sized current cars. But

---

20. Amann, op.cit.
21. The higher costs of diesel cars have not deterred their purchase in countries such as Italy which have low taxation on diesel and hence a low retail price.
22. See Bleviss, op.cit.

Figure 4.2 Model-based calculation of the impact of improving efficiency, for given base vehicle characteristics

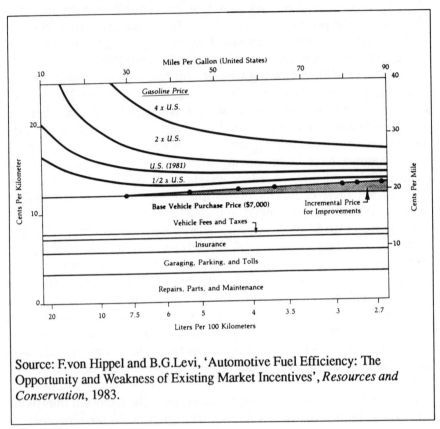

Source: F.von Hippel and B.G.Levi, 'Automotive Fuel Efficiency: The Opportunity and Weakness of Existing Market Incentives', *Resources and Conservation*, 1983.

Figure 4.2 illustrates the key point about efficiency-cost trade-offs. Life-cycle costs can vary relatively little across a very wide range of efficiencies, particularly for low fuel prices. At the highest fuel prices illustrated, lifetime costs fall with increasing efficiency across the full range shown.

4.7 Market barriers

The motor car is a potent symbol of success and status. The advantages, or at least the perceived advantages, of freedom, privacy and convenience, means that there is an almost universal desire to own one.

Most people acquire a car as soon as they can afford one, and their choice of vehicle is more often associated with 'gut feelings' than with any rational decision, unless there is a clear cut personal advantage of a particular choice, such as lower capital cost. In Europe this reliance on gut feeling is illustrated by the opinion polls that have consistently shown that the public is concerned about the environment and that in general people are prepared to spend more on an 'environmentally friendly' product, but when it comes to buying a new car consumers prefer to spend their money on luxury items such as electric sun roofs than on emissions control devices, unless there is a direct financial advantage such as a tax incentive.[23]

Consumer interest in fuel efficient technology has followed fluctuations in oil prices. Following sharp price rises, consumers tend to demand fuel efficient cars, and when prices slacken demand focuses on performance, style, accessories, and other vehicle characteristics. Fuel prices at the end of the 1980s were lower in real terms than before the first oil crisis in 1973, and for several years there has been little consumer encouragement for manufacturers to produce energy efficient products. In addition, one of the results of the improvements in energy efficiency in the late 1970s and early 1980s has been that fuel costs are now a smaller proportion of total operating costs and therefore an individual has less to gain from purchasing a highly energy efficient vehicle. Typically fuel prices represent only about one-sixth of the cost of buying, operating and maintaining a modern passenger car in the UK, and less in other countries where fuel prices are lower.

The desire to reduce motoring costs is unlikely to motivate consumers to alter significantly their preference for makes and models of automobiles at current fuel prices. It has been widely argued that fuel prices should be raised through increased taxation to penalize those who use inefficient vehicles and do high mileages, and to encourage the purchase of more efficient technology. However, as illustrated in Figure 4.2, even with doubled fuel prices the total costs vary little over a very

---

23. In West Germany, where tax incentives have been given for the purchase of cleaner cars, 97% of all cars sold in 1989 were equipped with emissions control devices. In the UK, where there was no tax incentive, less than 1% of new cars sold were equipped with the devices in the same year.

wide range of efficiencies.[24] With so many other factors helping to determine consumer preference, the financial incentive does not register strongly enough to create significant demand for greater efficiency. Even those who are concerned about costs may well focus more on the up-front capital costs than long-term running costs. This explains why opportunities exist for improving efficiency at little cost, and why petrol prices would have to increase substantially for a significant change in consumers' buying habits, and remain at a high level relative to incomes over a long period of time to maintain an upward trend. It is unlikely that sufficiently large price hikes would be politically acceptable in western democracies, though that is not to say that increased taxation does not have a role to play in reducing energy consumption from road traffic.

4.8 Policy options

Policy can focus on ways of encouraging producers to make vehicles with advanced energy efficient technology, and on consumers to buy more efficient vehicles.

Producer regulation and incentives

In the past, many industrialized countries adopted voluntary fuel economy targets for their vehicle industries. The only country to adopt a mandatory target was the USA, where the Corporate Average Fuel Economy (CAFE) regulation was agreed in 1975, which set a timetable for the introduction of more fuel efficient cars from 1979. These standards were based on the average consumption of all cars from a given manufacturer (or importer), and if the company failed to meet the target, a fine of US$5 for each 0.1mpg over the CAFE target for every car sold had to be paid to the government.

A major criticism of the regulation was that the structure made it much easier for manufacturers selling mainly small cars (ie. the Japanese) to meet the standards than those selling both large and small cars (namely the US car makers). Those selling solely large cars suffered most (namely the European luxury car makers). To make the target fairer in future, many have called for the CAFE to be set on a percentage improvement basis, which would require the same degree of effort by all

---

24. Figure 4.2 is from a US study, so the impact would be greater in most other countries which have higher fuel prices, but the point remains valid.

manufacturers. This, on the other hand, would penalize those manufacturers, particularly the Japanese, that have already adopted advanced technology to improved their products' fuel economy. Out of the big three US car manufacturers only Chrysler supports the retention of the CAFE regulations,[25] the others favouring increasing fuel prices for promoting more efficient technologies.

Another modification of the CAFE regulations has been submitted informally by the British government to the European Community.[26] Under the proposed scheme tradeable credits would be used to ensure compliance with a corporate average fuel economy standard. A manufacturer that exceeds the fuel economy standard earns credits which can be sold to those which fail to meet it, primarily the manufacturers of large cars. As the standard becomes more stringent, market forces will increase the cost of the credits, and thus the manufacturers producing predominantly large cars will be forced to subsidize those producing small cars.

Another approach is for governments to give incentives to manufacturers along the lines of the American Alternative Motor Fuels Act. This provides car manufacturers with incentives to produce vehicles capable of running on natural gas, pure alcohol or a mixture of alcohol and petrol.

In Europe the car manufacturers would prefer an extension of the type of voluntary agreement reached in the late 1970s, whereby the industry agreed to improve fuel economy by 10% between 1978 and 1985. It has been argued that the industry set a target which it knew would be easy to meet and that the agreement had no technology forcing effect on the manufacturers in Europe. In the event, the improvements were over 20%, and many consider that market forces, at times of high fuel prices, actually had a greater influence on the manufacturers. The European motor industry has in late 1991 voluntarily committed itself to a 10% reduction in CO_2 emissions from new cars (equivalent to a 10% reduction in fuel consumption) over the period from 1993 to the year 2005.[27] Also under discussion in the European Community is the imposition of

25. A.Flax, 'Chrysler: CAFE Hike Possible', *Automotive News*, 8 May 1989.
26. 'Rifkind calls for urgent action on fuel economy: UK puts tradeable permits proposal to EC', Press Notice No 202, Department of Transport, 9 July 1991.
27. Paper presented by Association des Constructeurs Europeens d'Automobiles, to Motor Vehicles Emissions Group of the European Commission.

carbon dioxide emissions limits which would, in effect, require the adoption of some route to limiting fuel consumption.

In the longer term, manufacturers need to be encouraged to develop new energy efficient technologies through increased government research and development funding targeted to joint industry/government programmes. The Japanese car industry, which has been particularly successful in developing advanced technology, receives considerable support for its long-term research and development programmes, and each company - unlike their European and American counterparts - does not try to develop every possible lead, but instead specializes in particular areas.

Consumer incentives

One of the main objections to the US mandatory CAFE regulations was that the manufacturers were forced to produce cars that the consumer did not want. Therefore, coupled with an encouragement for industry to produce fuel efficient vehicles, there must be an incentive for consumers to value efficiency more than power and performance. It has been suggested that price increases, through the use of 'carbon' taxes or 'energy' taxes, would return consumer interest to fuel consumption, but as outlined above this on its own is unlikely to be sufficient to encourage the purchase of the optimal energy efficient technology. In those countries where fuel is subsidized, presumably to promote development, it may be appropriate to remove such subsidy to encourage energy conservation.

Other options to encourage consumers to buy energy efficient vehicles include the adoption of a 'gas guzzler' tax as was introduced in the US in 1980 to discourage people from buying very inefficient vehicles. It is a progressive tax, starting with cars having a fuel consumption above 10.5 litres/100km. There is some evidence that the tax has provided a major incentive to car manufacturers to design their vehicles to avoid this tax.

Another way of stimulating consumer demand for low energy vehicles is by offering a tax incentive to people buying these vehicles, in a similar way to those offered in some European countries for vehicles fitted with emissions control devices before they become mandatory in 1993. In California, it has been proposed that tax incentives for both cleaner and more energy efficient vehicles should be offered. Under this scheme, called DRIVE+, it is proposed that sales surcharges on vehicles with high emissions (including carbon dioxide) would pay for the tax deduction on the more efficient, low-emitting vehicles and the costs of administering the scheme.[28] A national version of DRIVE+, known as a 'gas guzzler tax; gas sipper rebate' scheme has been proposed for the US that involves expanding the gas guzzler tax to collect money from the manufacturers of inefficient and polluting cars and give it to the purchasers of efficient and clean vehicles. It has also been proposed in Massachusetts that the purchase tax on new cars should vary depending on the efficiency of that vehicle compared with others in its size class. The least efficient vehicles within a given size class would pay a 10% tax and the most efficient would pay none. The average tax per vehicle would be at the current level of 5% so that the tax revenues remain unchanged.[29]

Finally, to help change attitudes towards the relative merits of high performance verses fuel economy governments could impose restrictions on advertizing, in order to promote fuel economy rather than performance, speed and acceleration. The UK government, for example, is considering encouraging the development of new guidelines on advertising practice which would encourage car manufacturers to lay less stress on performance and more on efficiency and safety.[30]

Regional issues

Western Europe, North America and Japan are the world's main car markets, accounting for three out of every four cars in use around the world, and are likely to continue to be so for the foreseeable future despite growing demand elsewhere because these countries already have a large number of vehicles, many of which are replaced each year. However, the largest potential market lies in the non-OECD countries,[31] particularly

---

28. D.Gordon and L.Levenson, *DRIVE+: A Proposal for California to Use Consumer Fees and Rebates to Reduce New Motor Vehicle Emissions and Fuel Consumption,* Applied Science Division, Lawrence Berkeley Laboratory, Berkeley, California, 1989.
29. J.Koomey and A.H.Rosenfeld, 'Revenue-Neutral Incentives for Efficiency and Environmental Quality', *Contemporary Policy Issues,* Vol. VIII, July 1990.
30. UK Department of Transport, Evidence to House of Commons Energy Committee, Inquiry into Energy Efficiency, HMSO, 1990.
31. In Western Europe car ownership is around 400 cars per thousand population; Japan 270; Australia 450 and USA 625. Most eastern and central European countries have car ownership levels of less than 200 per thousand population; in the USSR it is less than 50 (MacKenzie and Walsh, op.cit.).

those of eastern and central Europe and parts of Asia. Western motor manufacturers have already begun establishing manufacturing and distribution networks in eastern Europe and the size of the potential market in Asia is perhaps illustrated by the rapid increase in new car and truck sales in Thailand which in 1990 alone increased 40% from the previous year, albeit from a low base.[32]

Western Europe, North America and Japan also dominate car production, although in recent years other countries, such as the former USSR, Korea and Brazil, have emerged as significant producers. Vehicle specifications in these three main markets/producer countries will, to a large extent, determine those of the rest of the world, as manufacturers aim for global harmonization and new producer countries have few resources to develop their own standards. For example, stringent energy efficiency requirements in these three dominant markets could lead to similar requirements elsewhere, especially in areas where vehicle production is increasing and manufacturers are looking to export to Europe and North America. In addition, the improved fuel economy of western vehicles will be reflected in both new and second-hand vehicles exported for use in developing countries.

Given the rapid growth in non-OECD countries, improving fuel economy of new cars sold outside the OECD will become as important as it is for the industrialized. It is vital that these countries are not seen as new markets for selling old technology, as has happened in the past. One way round this is to ensure that models are not exported unless they meet the same standards as those for the domestic market. Alternatively, and probably more effectively, importer countries could establish purchase taxes based on the fuel consumption of the vehicles to encourage producers to sell the most energy efficient technology. As noted, there is a higher proportion of old cars in non-OECD countries and turnover is slower. To encourage more rapid efficiency improvements, financial incentives could be offered to owners who scrap their old inefficient cars, as in Greece, and policies such as those discussed above for new cars could be considered for second-hand imports.

---

32. Michael Dunne, 'Thailand Sales in Pleasant Rut - Up 40% a Year', *Automotive News*, 18 February 1991. Car ownership levels in Thailand are about 10 per thousand population; in Singapore 100; in Korea 37, but in India it is 2 and China less than 1 car per thousand population (MacKenzie and Walsh, op.cit.).

The US and European motor industries are currently (1991) in a period of recession and profit margins have been cut to a minimum, especially in North America, discouraging innovation towards more efficient technologies. However, if the companies in these areas do not face the new challenge of fuel efficiency they could lose further ground to the Japanese, and, increasingly, manufacturers of industrializing countries such as South Korea.

4.9 Prospects

It has been estimated that proven technology, that is technology that either is already used in certain production models or that included in manufacturers' plans for future production lines, could reduce new car fuel consumption in the US by up to nearly 50% by the end of this decade, at an additional vehicle cost of less than 10%.[33] Other analyses, by the US Office of Technology Assessment and the US Department of Energy, suggest that improvements approaching 20% in new fleet averages could be achieved by 1995 (based on the 1987 model year).[34]

In Europe there is likely to be similar potential despite already having more energy efficient cars on the market. An analysis by the UK Energy Technology Support Unit suggests that the current best available technology offers a fuel consumption improvement of 20-30% while a French analysis suggests the potential lies in the range of 40-50% for medium and large cars, and 25-30% for small cars.[35]

However, motor manufacturers on both continents view these improvements as hugely over-optimistic. The industry has given little indication of the maximum technical potential for improved fuel economy, but it is known that they are currently forecasting, assuming a 'no change' scenario, very small improvements in sales weighted new car fuel consumption. For example, in the UK improvements of less than 5% by the year 2000 have been forecast by the industry.[36]

While the independent estimates of the potential for improvements in energy efficiency are possibly over-estimated because some fail to take

33. C.Difiglio, K.G.Duleep and D.L.Greene, 'Cost Effectiveness of Future Fuel Economy Improvements', *The Energy Journal*, Vol.11, No.1, 1990.
34. *Automotive News*, 8 May 1989.
35. Papers presented Berlin conference, op. cit., footnote 1.
36. 'The Motor Industry and the Greenhouse Effect', Society of Motor Manufacturers and Traders, London, 1990.

account of the more stringent emissions controls that have been adopted since the analyses were undertaken, there is a large untapped pool of proven technologies that could be introduced to more than compensate for any small energy penalty. Difiglio et al,[37] for example, have looked at the potential for reducing new car fuel economy using technology that is already in manufacturers' product plans for 1995. Assuming that vehicle performance, size and accessories stay at the same level as in 1988, this analysis suggests that the fuel economy of new cars could improve by nearly 20%.

It is obvious that market forces at times of low energy costs will only drive very small improvements in energy efficiency. Without some form of government intervention the available technology will be channelled into making vehicles more powerful rather than saving energy. The extent to which the potential savings are realized will depend upon the development of policies as discussed above.

4.10 Conclusions

Available technology could deliver between 20-40% improvements in average fuel economy for new cars over the next decade or so, despite the introduction of more stringent emissions controls at the same time. This will only occur with government intervention in the form of a package of measures, particularly financial incentives, to both the manufacturers and the purchasers of cars. Left to market forces alone energy efficiency improvements will be small if demand for yet higher performance, and more luxury energy-consuming accessories continues.

Increased fuel taxation has an important role to play in reducing energy consumption from the transport sector. However, to stimulate sufficient consumer interest to exploit most of the technical potential fuel prices would have to rise steeply and rapidly, as they did during the oil price crises of the 1970s. It is unlikely that such prices rises would be politically acceptable.

One of the most effective ways of encouraging the industry to produce the most energy efficient cars is through a modified CAFE scheme. However, for this approach to be effective it must be accompanied by financial incentives for the consumer to buy low energy cars such as a 'gas guzzler tax; gas sipper rebate' system. This dual approach of 'pushing' industry and 'pulling' consumers is likely to be more effective

37. Difiglio et al, op.cit.

than other measures at introducing advanced energy efficient technology onto the market in the short term.

Currently, the OECD countries dominate the world car market, but this will not always be the case. In coming years eastern and central Europe and certain parts of Asia, where new car sales are increasing at a phenomenal rate, will become increasingly important. It is vital that these areas are not dumping grounds for old unsophisticated technology, but instead sell cars that are as energy efficient as those in the OECD countries.

Energy Savings in Domestic Electrical Appliances

There are substantial opportunities for saving electricity by improvements in the design of major household appliances, or white goods. White goods account for over 30% of total domestic electricity used in the UK (second only to heating), equivalent to more than the output of two large (2,000MW) coal-fired power stations, or about 14% of all UK electricity consumption.

Refrigeration appliances can be designed to use one-third the electricity of even the best models currently available in the UK. Such performance is attained by appliances which have been built for high efficiency and long life without consideration of purchase cost, or are as yet in the development stage. Models are commercially available in mainland Europe which have a very much better performance than any available in the UK. The intense competition in the High Street, in which purchase price, not lifetime cost is the main consideration, is an important barrier to the availability of high performance appliances. However, lowest first cost does not necessarily equate with minimum performance. Efficient models can be significantly cheaper than less efficient appliances of the same size.

Improved standards could possibly come about by greater public awareness of the benefits of higher efficiency. This could be brought about partly by a labelling scheme giving the lifetime and running costs of appliances, as in the US. This is mandatory in the US, as are minimum performance standards, based on the requirement for a maximum payback of 3 years. The UK government is now advocating a common

European energy labelling scheme, and minimum efficiency standards,
but at present favours a voluntary approach.

Developing countries could also benefit from improvements in appliance
efficiency. Substantial reductions in electricity demand would defer the
need for investment in generating plant, as well as saving expensive
imported fuel and benefiting consumers.

This chapter examines the potential for efficiency improvement in domestic white goods, concentrating mainly on refrigeration. It first looks at current electricity consumption and potential for improvement. It then looks at the range of new models available in the UK, and estimates the savings potential if consumers bought the most efficient. Finally it briefly reviews the US appliance efficiency standards for refrigeration, and examines their relevance to UK conditions.

5.1 Electricity use in domestic appliances

Every year a total of about £20 billion is spent by residents of the European Community on domestic electricity consumption. Of this amount, over 30% goes on running 'white goods' (cookers, refrigerators, freezers, washing machines, dryers and dishwashers); in homes without electric (space and water) heating, such white goods usually account for more than half of electricity consumption. The most important of these in terms of electricity use are refrigerators, fridge-freezers and freezers, collectively called refrigeration appliances. They cost £3-4 billion a year to run. A report for the EC showed that there is a potential for saving 15-25% of European domestic electricity use.[1] In white goods the potential is higher at 25-40%, and the greatest potential, at nearly 60%, is in freezers.

Expenditure on electricity for domestic appliances in the UK exceeds that of industrial motors, and is over twice that on commercial lighting. White goods cost nearly £2,500 million a year to run, with refrigeration costing £1,200 million (see Figure 5.1). The electricity consumption of domestic white goods takes about half the output of all UK nuclear power

---

1. 'The potential for energy saving in the applications of electrical energy', Fichtner study for the Commission of the European Communities, Fichtner, Stuttgart, June 1988.

Figure 5.1 Estimated consumption and running costs for UK white goods

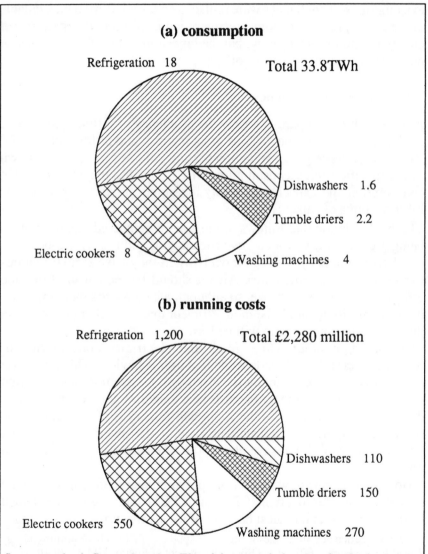

(a) consumption

Refrigeration 18   Total 33.8TWh

Dishwashers 1.6

Tumble driers 2.2

Electric cookers 8

Washing machines 4

(b) running costs

Refrigeration 1,200   Total £2,280 million

Dishwashers 110

Tumble driers 150

Electric cookers 550

Washing machines 270

Source: author's figures based on Electricity Association data for England and Wales consumption, and domestic unrestricted price in 1990 of 6.7p/kWh. The total UK electricity consumption in 1990 was about 265TWh (*Digest of UK Energy Statistics*, HMSO, 1991).

stations, or more than two large (2,000MW) coal stations. A recent report by March Consulting Group, for the Energy Efficiency Office, found that a saving of 40% (worth £1,500 million a year) could be achieved if the existing stock of appliances were replaced by the most efficient available in the UK.[2] In the UK home, refrigeration is the second largest user of electricity after space heating, but the largest in terms of cost because it does not benefit from cheaper off-peak rates.

5.2 Refrigeration efficiency

It is difficult to compare precisely the efficiency of different models of refrigeration because each varies in its features and utility, but a useful indicator is the energy consumption per unit volume (or capacity) known as the specific energy consumption (SEC). This has the advantage that consumption and capacity are easily measured. It can be expressed as kWh per litres of capacity.

In the discussion that follows, it must be remembered that the SEC is being used as a proxy for energy efficiency[3] and is only a good guide to relative efficiency between models when they have similar features (which is generally the case). Also it should be remembered that the efficiency improves with size, so although a larger-sized model will use more electricity it should be more efficient than a smaller one, because the surface-to-volume ratio will be less.

Compared to other white goods the electricity consumption of refrigeration is relatively independent of user behaviour. This is because less than 20% of consumption is caused by door openings and insertion of food. Thus the dominant factor is heat gain through the walls, which depends on temperature in the room and thickness of insulation. The former is important but frequently ignored: lowering the kitchen temperature by 5°C will save about 30% of electricity use for a typical refrigerator.

Most insulation is now made from polyurethane foam expanded with the chlorinated fluorocarbon (CFC) gas, CFC11. This gas is to be banned by the year 2000 at the latest because of its adverse impact on the ozone

2. 'Energy Efficiency in Domestic Electric Appliances', March Consulting Group Report for the Energy Efficiency Office, Department of Energy, HMSO, 1990.
3. From the definition, SEC is *inversely* related to energy efficiency ie. a *lower* SEC is a higher energy efficiency.

layer.[4] Other gases for expanding foam insulation can be used, but they have higher thermal conductivity, which would result in greater electricity use unless wall thickness is increased. An alternative is to use vacuum insulation, either using powder or solid spacers to keep the walls apart. Such refrigerators are being developed in Japan.

The compressor system generally uses other CFC gases, CFC12 and CFC22, as refrigerants. CFC12 is similar to CFC11 in its ozone depleting properties, and will have to be replaced by substitutes. What effect this will have on compressor efficiency is still unknown. However efficiency can be improved by 15% by using rotary compressors instead of the standard reciprocal type.[5]

5.3 Performance variation and cost: an illustration from the UK

The efficiency of models currently available in the UK is derived from a database of over 150 models compiled from information contained in *Which* magazine, Electricity Board shops and John Lewis department stores. The database, which was compiled by the author, contains details of capacity for both fresh and frozen food, electricity consumption and purchase cost.

From this database it can be established that:

* there is no consistent relationship between efficiency and unit cost;

* more efficient models are often cheaper than less efficient models;

* in the most popular size ranges efficiency can vary by a factor of 3, with up to £25 a year difference in running costs, between least and most efficient;

* the extra capital cost (corrected for differing sizes) of buying the most efficient models compared to the average varies from nothing for upright freezers, to £10 for refrigerators and chest freezers, and £25 for fridge freezers. Payback time for these higher capital costs is less than three years;

---

4. Under the 1990 London Revisions of the Montreal Protocol, CFCs are to be phased out by the year 2000, and a number of countries have announced intentions to phase out all CFCs before that.
5. The discussion in this section is based on Jorgen Norgard, 'Low Electricity Appliances: Options for the Future', in T.Johansson et al, *Electricity: Efficient End Use and New Generation Technologies, and their Planning Implications*, Lund University Press, Lund, Sweden, 1989, pp.132-9.

* in general, the most efficient model are made in Germany and Denmark, and the least efficient in the UK.

Table 5.1 shows size (capacity) in litres, average unit consumption (AC) in kWh and SEC in kWh/litre, for a range of models; the efficiency data is illustrated in simplified form in Figure 5.2. The 'stock' model is the average of all existing models currently in homes in the UK, and values given are those estimated by the CEGB.[6] The 'new' model is the 'sales-weighted' average (ie. the average in each size band weighted by the popularity of each size band) and is composed of models which are currently available in the UK. The 'best' is the model with the lowest SEC in the most popular size range.

Refrigerators with storage (freezing compartment) have a higher electricity consumption than those without (larder refrigerators). The most efficient refrigerators with storage have an SEC about 10% lower than new. The best larder refrigerator has an SEC 35% lower than average new. Combining both types in proportion to sales, gives an SEC for new the same as stock, but best about 20% less than stock.

There are more efficient larder refrigerators available in Denmark. For instance the Gram range includes a 200 litre model using only 90kWh a year.[7] Thus its SEC is only 0.45 compared to 1.14 for the best Gram model currently available in the UK.

For fridge/freezers, the electricity consumption of new models is only slightly less than stock, perhaps due to the influence of 'frost-free models'. The best (two) models available in the UK have half the SEC of existing (stock) models, and are both made in West Germany. More efficient models could be developed: a 500 litre prototype from the Technical University of Denmark has an SEC of only 0.4, or a third of the best available in the UK.[8]

Upright freezers have a higher SEC than chest freezers, because of their vertical door opening. New models use a 37% less electricity than stock, but they are 24% smaller, so the SEC is only 17% less. The best model has an SEC about 25% less than stock.

Chest freezers are more efficient than upright freezers since heat gain during opening is much less. New models use 30% less electricity, but they are 30% smaller, so the SEC is similar. The best model (from

6. *Planning for Hinkley Point C: Proof of Evidence CEGB Vol.4*, HMSO, 1988.
7. March, op.cit.
8. Norgard, op.cit.

Table 5.1 Capacity, consumption and efficiency of stock, new and high efficiency refrigeration available in the UK

Type	Capacity (litres)	Annual consumption kWh	Efficiency kWh/litre (SEC)	Index
Refrigerators				
Stock	145	300	2.07	100
New				
With storage	144	325	2.26	109
Larder	141	260	1.84	89
All	143	295	2.06	100
Best				
With storage	140	280	2.00	97
Larder	175	200	1.14	55
All	155	250	1.61	78
Fridge/freezers				
Stock	285	725	2.54	100
New	240	580	2.42	95
Best	255	310	1.22	48
Upright freezers				
Stock	205	760	3.71	100
New	155	480	3.10	84
Best	165	455	2.76	74
Chest freezers				
Stock	275	680	2.47	100
New	195	465	2.38	96
Best	275	310	1.13	46

Source: author's estimates from his database.

Denmark) is the same size as stock but uses less than half the electricity consumption. In West Germany, there are still more efficient models, with SEC 60% that of best model in the UK. A prototype model from

Figure 5.2 Efficiency of refrigeration available in the UK

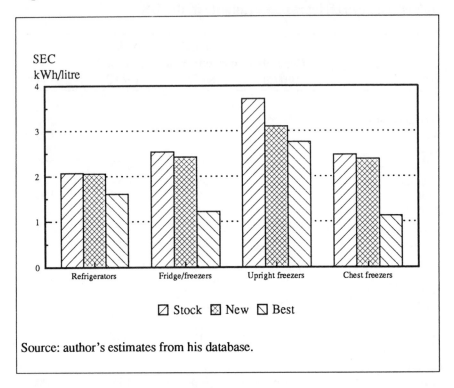

Source: author's estimates from his database.

Technical University of Denmark has an SEC a third that of the best UK model, and only a sixth of UK stock.

If consumers chose the most efficient model available (in the most popular size ranges) which did not cost any more than the average 'new' (termed 'no extra cost' models), then (correcting for any differences in size) the realizable savings potential would be about 3.5TWh, or two-thirds of the savings from buying the most efficient regardless of cost (the 'best').

The total extra cost of buying the best compared to average current models in shops is about £450 million and annual savings are about £370 million, giving a simple payback time for the greater capital costs of less than two years. Payback times vary from 2-3 years for larder refrigerators, to 1-2 years for fridge/freezers and refrigerators with storage, to less than a year for chest freezers. If consumers bought the 'no extra cost' models then annual savings would be £250 million at no extra cost. The marginal return from buying the best compared to the 'no extra cost' models is less than four years; overall a saving of £120 million a year for total expenditure of £450 million.

This demonstrates the importance of consumers knowing the running costs of models, and being able to easily calculate the payback from purchasing the more efficient models.

5.4 Appliance standards

The US National Appliance Energy Conservation Act of 1987 sets energy consumption standards for 12 groups of consumer appliances, including refrigerators, fridge/freezers and freezers.[9] Most of the standards were scheduled to go into effect on 1 January 1990, although some are not scheduled to take effect until 1992 or 1993, while others will become effective on 1 January 1998. The drive for this legislation was political, stemming from the energy-conscious attitude of the Carter administration.

To support these standards the US Department of Energy (DOE) is required to maintain uniform testing procedures and periodically reassess the standards and amend them as necessary. For refrigeration, the law states a specific schedule for re-analysis of the standards, and an amended standard from 1 January 1993. Under the law DOE is required to determine whether an amendment of the standard achieves 'the maximum improvement in energy efficiency that the Secretary determines is technologically feasible and economically justified'. The law specifies a maximum of a three year payback to consumers on the additional cost of efficiency improvement.

A potential problem in applying the US refrigeration standards to UK models is the fact that US models are at least twice as large (in volume) as UK ones. For instance a typical fridge/freezer has a volume of 400-600 litres in US, compared to 250-300 litres in the UK. Similarly a typical chest or upright freezer has volume of over 600 litres in US compared to less than 300 in the UK. Thus it is necessary to demonstrate that the observed relationship between energy consumption and volume for US-sized equipment holds for the much smaller UK equipment. Preliminary results from the Lawrence Berkeley Laboratory (who

9. Energy Conservation Program for Consumer Products; Proposed Rules and Notice, US DOE Office of Conservation and Renewable Energy 10CFR Part 430 in Federal Register Vol.53 No.232, 2 December 1988.

Table 5.2 Annual consumption of UK 'best' refrigeration relative to US standards (kWh)

| | UK equipment | | US standards | |
	Stock	Best	1990	1993
Refrigerator	300	235	405	325
Fridge/freezer	725	350	502	405
Upright freezer	760	565	560	335
Chest freezer	680	310	475	355

Source: compiled from data contained in Table 5.1

perform the analytic work for the DOE) indicate that standards would be valid for the smaller-sized UK models.[10]

The US standards give consumption in kWh as a function of adjusted volume which takes into account the extent of the freezing compartment. Table 5.2 shows the results of applying these US standards to UK refrigeration for models with capacity of UK stock, which is assumed to be manual defrost type. The consumption for 'best' is UK stock capacity times 'best' SEC from Table 5.1.

For refrigerators, application of US 1990 standards would not reduce consumption of stock. For fridge/freezers and freezers application of 1990 standards would reduce consumption by 25-30%, the maximum proposed 1993 standards by 45-55%. The best models available in the UK meet 1990 standards, with best fridge/freezers and chest freezers easily meeting maximum proposed US 1993 standards.

In conclusion the application of the US appliance standards to UK refrigeration equipment would, if applicable to the much smaller models found in the UK, substantially reduce energy consumption. Except for

10. Private communication from Isaac Turiel (Lawrence Berkely Laboratories, US) to the author.

upright freezers, models are available in the UK which meet, and exceed, the 1990 and proposed 1993 standard.

The US standards for 1993 fall far short of the best already demonstrated. For example, the SEC of 1993 refrigerators in the above table is only 2.24, compared with 0.45 for the Danish Gram refrigerator referred to in section 5.4. One reason why the standards are not tighter may be due to the requirement that any increased capital cost resulting from the standard must be recoverable from the electricity savings in less than three years (at the relatively low US electricity prices).

5.5 Potential for energy savings

If consumers bought the most efficient refrigeration now available in the UK (with same size), electricity consumption for refrigeration could decline by about 9TWh, or 50% of current consumption. The savings potential in the future depends, however, upon future ownership levels and consumer choice of models. Thus the realizable potential in the year 2000 will be less than current potential because some of efficiency gains will have been taken up, as consumers replace their current models with newer, more efficient models (a process referred to as 'stock turnover'). There is also a tendency towards smaller models.

Table 5.3 below shows estimated annual consumption levels for 1990 and projections for year 2000 for 'new' and 'best' models. The realizable saving is the difference between best and average. Total potential is 5TWh, or 35% of projected consumption in the year 2000. The majority, 60%, of the savings comes from fridge/freezers, and about 15% each from refrigerators and chest freezers.

If there had been no efficiency improvements between now and year 2000, then total consumption would have been 19.2TWh, instead of 15.7TWh. Nevertheless 'natural' stock turnover will produce 3.5TWh savings, or 40% of current savings. Thus there still exists a large realizable potential for savings, worth over £350 million a year at current prices.

5.6 Energy labelling: the UK debate

Energy labelling is a scheme whereby consumers are given information on the running cost of an appliance before purchase (like an mpg label on a car). It can be in the form of a sticker attached to the appliance which

Table 5.3 Estimated energy consumption for refrigeration in 1990 and 2000

	Annual consumption (kWh)			Total UK consumption (TWh)			Energy savings (TWh)
	1990 stock	2000 average	2000 best	1990 stock	2000 average	2000 best	
Refrigerators	300	295	230	3.5	3.7	2.9	0.8
Fridge/freezers	725	580	295	8.1	6.9	3.5	3.4
Upright freezers	760	480	430	3.4	3.3	3.0	0.3
Chest freezers	680	465	220	2.7	1.7	0.8	0.9
Total				17.7	15.6	10.2	5.4

Note: consumption of 'best' is capacity of 'new' times 'best' SEC taken from Table 5.1. Appliance ownership projections were derived from the 1988 CEGB forecast, whose data are for England and Wales.

Source: author's calculations

gives its running cost in comparison to lowest and highest; or at minimum a leaflet explaining how to calculate the running cost of an appliance.

The former approach is taken in the USA where 'Energy Guide' labels are required by Federal law on major appliances (including refrigerators, fridge/freezers, freezers).[11] The label shows the estimated annual running cost of the appliance, together with the lowest and highest running cost of models in that size range. Labelling has been shown to have clear impact, not only in the US, but for example in New South Wales where mandatory labelling resulted in an average 15%

11. *Appliance Labelling*, leaflet issued by US DOE Department of Conservation and Renewable Energy, DOE, Washington.

improvement in the efficiency of refrigerators sold, and models with poor energy ratings nearly disappeared.[12]

The UK government has powers to require energy labelling under the 1981 Energy Conservation Act. Labelling could be bought in at any time by statutory instrument, that is, without going back to Parliament for approval of new legislation.[13] The only action was in 1985 when the Energy Efficiency Office gave out a leaflet, explaining how to calculate the cost of running refrigeration appliances. However after a short while the EEO stopped printing the leaflet, and has not given out any further information. In April 1985 the Eastern Electricity Board attached energy cost labels to refrigerators and freezers displayed in its 122 shops.[14] They reported a relative lack of interest in energy efficiency as a cost issue, though about a seventh (13%) of appliance buyers claimed that the labels at point of sale had influenced their choice.[15]

The Energy Efficiency Office encourages voluntary labelling, but the UK manufacturers' view is that British goods would be placed at a disadvantage if energy cost data were not available for imported goods. Also there is concern about the cost of an energy labelling scheme, which was estimated at about £16 million in 1980.[16] This, together with lack of support from UK manufacturers and retailers has stopped any labelling or information scheme being undertaken, but given the large savings available to consumers from informed purchases (over £350 million a year just from refrigeration appliances), the cost - even if borne by consumers - of a labelling scheme would be rapidly repaid even if only 10% of consumers paid attention to it.

Up until 1990 the UK government was hostile to energy labelling, describing it as an 'unnecessary bureaucracy which would prove tedious and turn people against energy efficiency'.[17] However in its Environment White Paper, published in September 1990, the government states that it would press '...for agreement in the European Community on a common energy labelling scheme, and minimum

12. 'Signals from Brussels', *Financial Times*, 16 October 1991.
13. Brian Brinkworth, 'Energy Labelling', *Energy World*, August 1989.
14. National Audit Office, *National Energy Efficiency*, HMSO, July 1989, para 3.16-3.19.
15. March, op.cit.
16. Department of Energy Consultative Document, DoEn, March 1980.
17. Reply by Mr. Morrison, Minister of State for Energy, to Parliamentary Question, 27 July 1989.

efficiency standards for equipment such as central heating boilers, fridges, washing machines and industrial heating'.[18]

Later in the White Paper[19] the government states that it 'favours a voluntary approach in the first instance', and that it will 'encourage and help the Commission ... to devise a workable scheme of the kind that manufacturers would be willing and able to comply with.' However the National Audit Office in its recent report on national energy efficiency stated:

> ...without the stimulus of compulsory labelling, along with consumer comparisons of the energy efficiency of competing appliances in purchasing decisions ... there is reduced incentive upon appliance manufacturers to improve the energy efficiency of their products.[20]

Government action has not however been encouraging: it blocked a proposal by the European Commission for legally-binding standards on domestic appliances, and now plans to label only the more efficient appliances.[21] Voluntary agreement with manufacturers seems as difficult as ever to reach, with a leading UK trade association requesting more time, a plea the *Financial Times* reported 'it also used to obstruct a similar initiative in 1979'.[22]

What is important is not the cost of binding standards or labelling schemes, but their cost effectiveness considered from the national point of view. It is not clear why there is reluctance to implement labelling schemes. Perhaps the reason is that the manufacturers of less efficient appliances do not wish inferior performance to be publicized.

5.7 Effect on manufacturers

In 1990 nearly half of the 2.7 million UK domestic refrigeration sales were of imports.[23] By far the most important exporter to the UK is Italy (Zanussi, Indesit and Philips brands), followed by Germany (Bosch, Liebherr and AEG brands) and east European countries (Yugoslavia,

---

18. *This Common Inheritance: Britain's Environmental Strategy*, HMSO, Sept. 1990, para. 5.31.
19. ibid, para. C32.
20. March, op.cit.
21. 'Britain Reneging on Pledges to Save Energy', *The Observer*, 11 Nov. 1990.
22. 'Black Mark for White Goods', *Financial Times*, 14 Dec. 1990.
23. 'Refrigerators and Freezers', *Economist Intelligence Unit Report*, No 406, EUI, December, 1991.

Hungary and the former USSR). In the UK the leading brands are LEC, Hotpoint and Tricity; their exports are less than 10% of domestic sales.

The UK brands tend to be concentrated on the smaller-sized, cheaper models mostly sold through Electricity Board shops, and high street chains like Curry, Comet and Rumbelow. These retailers account for over half of the £500 million market for domestic refrigeration. The main competitors to UK brands, particularly for refrigerators, are east European countries and Italy (Zanussi and Indesit brands).

March Group report that the largest European manufacturers do recognize that the UK market lags behind other European markets, especially Germany, in the demand for energy efficient appliances.[24] They manufacture more efficient appliances for other markets and are thus better able to introduce high efficiency models to the UK than British companies manufacturing mostly for the UK market. British manufacturers who do not make efficient appliances will be at a disadvantage, particularly in the Single European Market.

5.8 Other white goods

The UK white goods market is worth over £2 billion a year, with nearly 10 million appliances a year being sold. The most popular appliances are washing machines (2 million), microwaves (1.75 million), electric cookers (0.9 million) and fridge freezers (0.8 million). There are substantial opportunities to improve the energy efficiency of white goods. The March Group study on UK domestic appliances found potential energy savings of 25% in washing machines and dryers, 40% in dishwashers, 30% in cookers and 50-60% in refrigeration.[25]

The Fichtner study of European appliances estimated possible energy savings of 29-37% in washing machines, 9-25% in cookers, 24-32% in dryers and 35-50% in refrigeration.[26] A detailed study of appliances in Denmark found similar savings potential.[27] Its author estimated that, compared to current stock, electricity consumption in white goods in Denmark could be halved through use of best available, and could be

24. March, op.cit.
25. ibid.
26. Fichtner, op.cit.
27. Norgard, op.cit.

reduced by 70% from advanced technologies which were not yet commercially available but which were expected to be cost effective.

5.9 Relevance to developing countries

Appliance use is the second most important component, after lighting, of domestic electricity use in developing countries. The most important appliances are refrigerators, TVs, irons and air conditioners. A survey of domestic energy use in Indonesia found that there were substantial opportunities to improve the efficiency of local appliances.[28]

The survey in Java found that refrigerators were generally inefficient, and local products could be improved by 40%. Incandescent lighting could be changed to fluorescent bulbs, the efficiency of air conditioners could be improved by up to 40% and TVs by 20-25%. Overall a typical consumer could save at least 20% of electricity used. This figure is very similar to Fichtner's estimate of saving 15-25% in European domestic electricity consumption.

5.10 Conclusions

There is substantial evidence that domestic consumers could save up to 25% of their electricity consumption, both in the UK, in Europe and in some developing countries, *at net economic savings.* The major savings come in white goods, of which the most important in terms of electricity use are refrigerators, fridge/freezers, and freezers. *There are few if any significant trade-offs involved in using more efficient appliances, and the additional costs of the most efficient models currently available would be paid back in electricity savings generally within 2-4 years.*

Although the efficiency of refrigerators has improved by about 15% since 1975, and that of fridge/freezers and freezers by 30-35%, there are still major opportunities for improvement using existing technologies.[29] If UK domestic consumers replaced their existing refrigeration appliances with the most efficient available today then electricity consumption would fall by about 9TWh, or over 3% of total UK consumption. This saving would be worth about £600 million a year

28. Lee Schipper and Steven Myers, 'Improving Appliance Efficiency in Indonesia', *Energy Policy*, July-August 1991, pp.578-87.
29. D. Dossett, 'Future influences on energy consumption in household refrigerators', *Energy Management*, August 1989, pp.23-4.

at current prices, or about £25 per household. In the EC consumers have the potential to save £1-2 billion in running costs a year from purchasing the most efficient refrigeration appliances.

Consumers will not replace their existing models overnight, but by the year 2000 most will have been replaced. If consumers over the next decade buy the most efficient (the 'best') available rather than the average, then total UK annual savings would be over 5TWh in the year 2000, worth over £350 million a year. The extra cost of buying the most efficient compared to average would be about £450 million, giving a payback time, on average, of less than two years. Even buying the most efficient models which do not cost more than the average model (the 'no extra cost' model) would save nearly 4TWh, worth about £250 million annually, for no extra capital cost.

At the moment consumers do not have energy cost information and are thus not able to easily purchase the most cost-effective and efficient models. One solution is an energy labelling scheme as in the USA. There every model has its running cost information given in comparison to the lowest and highest cost model in its size range with similar features. This scheme has been law in the US since 1980.

The long-standing resistance to labelling schemes in the UK, despite the clear economic benefits of and rationale for labelling, illustrates some of the political obstacles in countries where ideological opposition to government intervention combines with the opposition from manufacturers that produce goods with efficiencies below that of many competitors abroad. But as the UK government admits, consumers need to be given information on the running cost of appliances so they can make the most cost effective and efficient purchase.

Standards could also be used to eliminate inefficient models. If the government, through British manufacturers' obstinacy, does not quickly introduce energy labelling it will undermine its credibility and leave the way open for the European Commission to act on consumers' behalf by introducing legally-binding standards on domestic appliances. The economic and environmental benefits available from more efficient appliances are too great to be left untapped.

Energy Efficient Lighting Technologies

Lighting accounts for about 15% of total electricity consumption in the developed countries. Recent advances in lighting technology, especially improvements in fluorescent lamps, could reduce projected energy consumption for lighting by 30-40% within a decade, with precedents for practice in developing countries. The saving would not entail any decline in lighting quality.

The two strategies with the greatest potential are a move from filament lamps to compact fluorescent lamps in the home, and in commercial applications from conventional fluorescent lighting to high frequency fluorescent lighting. Control systems to match the light output to the service requirements also offer considerable savings.

More efficient lighting technologies are of higher initial cost, but the extra cost is usually recouped by the electricity savings within the lifetime of the first lamp.

The reduction in lighting energy consumption depends on backing by governments for energy efficient lighting. Policy options include: legislation to make the adoption of energy efficient lighting mandatory where appropriate; various financing schemes including capital subsidies and leasing to overcome the first-cost barrier; and regulation of electricity companies requiring or encouraging them to promote energy efficient technologies including those for lighting.

6.1 Energy use for lighting

Lighting accounts for about 15% of electricity consumption in developed countries, and about 8% in developing countries, as shown in Table 6.1 opposite. The usage of the main types of lamp - filament lamps, fluorescent lamps and discharge lamps - in 1960 and 1990, and the quantity of light output (relative scale) for 1960 and 1990 are also shown. The table gives a broad-brush picture of the consumption of energy in lighting. Table 6.2 shows more detailed figures for the UK, where lighting accounts for 15% of electricity consumption.

In the developed countries, there has been a steady rise in lighting standards. One yardstick is the designed illumination level for office work. This rose from about 100 lux[1] pre-World War II to 500 lux by the 1970s. Since the energy crisis of the 1970s, there has been widespread agreement to freeze designed illumination levels, for example at 500 lux for office working areas. Research had shown this level to be adequate for most people for ordinary visual tasks in offices.

Growth in total light output has been greater than the growth of each national economy, due to up-dating of lighting during new building and refurbishment. In developing countries, the growth in use of light has been much faster than in industrialized countries because of their more rapid economic growth and low starting levels for electric lighting. However, the growth in energy used for lighting has been less than the growth in light output, and the share of energy for lighting in total energy consumption has been constant over the 1960-1990 period. This is due to the steady move to higher efficiency in lighting.

Table 6.3 shows, for the developed countries, the change from 1960 to 1990 in the share of energy between the different types of lamp.[2] The main trend has been the move from the low-efficiency filament lamp to the high-efficiency fluorescent lamp. The fluorescent lamp has also increased in efficiency (technically, efficacy[3]) from 75 lumens/Watt in 1960 to 100 lumens/Watt in 1990, mainly as a result of the introduction of new-technology phosphors (the fluorescent powder on the inside of

1. The total light output of a lamp is measured in lumens. Lux is a measure of light on a surface, in lumens per square metre of surface.
2. For a review of current lamp types see *Lamp Guide 1990*, Lighting Industry Federation. Figure 6.1, based on this Guide, shows the improvement over the years in the principal lamp types.
3. The efficacy of a lamp is the ratio of total light emitted by the lamp in lumens to the electrical input power in Watts.

Table 6.1 Growth of lighting in developed and developing countries

	Developed countries		Developing countries	
	1960	**1990**	**1960**	**1990**
Number purchased per capita per year				
Filament lamps	3	4.5	1	1
Fluorescent lamps	0.25	10	0.05	0.6
Discharge lamps	0	0.07	0	0.02
Light output (relative)				
of lamps above	100	750	15	250
Energy for lighting,				
(kWh per capita per year)	250	800	50	300
Lighting share of electricity				
consumption (%)	15	15	8	8

Source: 'Efficiency of Electricity Use', *House of Lords Paper 37*, HMSO 1989;
CIE Technical Committee TC 7-03.

the tube). With the latest lighting technology, the electricity cost of lighting for an office worker for one year is approximately equal to one hour's wages.

The trend to higher energy efficiency in lighting can be accelerated by the wider use of the latest lighting technologies. If these technologies are supported, significant savings in world use of energy can be achieved.

6.2 Saving energy in lighting

It is generally recognized that there are three principles which apply to saving energy in lighting:

* standards of good practice in lighting, eg. illumination levels as in the CIE (Commission Interationale d'Eclairage) recommendations, should not be reduced. Practices such as taking lamps out or fitting voltage-reducers which result in lower lighting levels are strongly

Table 6.2 Energy for lighting in the UK

	Consumption (TWh per year)
Lighting sector	
Commercial and public	20
Industrial	8
Domestic	7
Total lighting	35
Potential lighting saving (see section 6.10)	12-15
Total UK electrical energy	230

Notes: average power demand for lighting 5GW, requiring approx. 10GW peak generating capacity. 35TWh per year in the UK releases about 30 million tonnes of carbon dioxide and requires about 10 million tonnes of coal.

Source: 'Efficiency of Electricity Use', *House of Lords Paper 37*, HMSO 1989; CIE Technical Committee TC 7-03; 'Energy Policy Implication of the Greenhouse Effect', *Sixth Report of the Energy Committee*, House of Commons, Vol. 2, 192-II, HMSO, 1989.

deprecated. In the UK the principal authority for good practice is the Code for Interior Lighting of the Chartered Institution of Building Services Engineers;

* within the requirements of good practice, preference should be given to the more energy efficient lamps, ballasts and luminaires;[4]

* lighting should be controlled according to need - eg. the different lighting levels for different tasks, the presence or absence of people, the contribution from daylight.

4. These different components of a lighting system are: lamp - the replaceable lightsource, eg. filament bulb or fluorescent tubular lamp; ballast - the control gear for limiting the current of fluorescent and discharge lamps; luminaire - the technical term for the lamp housing, or lighting fitting.

Table 6.3 Share of lighting energy in developed countries, and lamp efficacies

	Share of lighting energy (developed countries)%	
	1960	1990
Filament lamps	67	27
Fluorescent lamps	30	55
Discharge lamps	3	18
Totals	100	100
	Efficacy lumens/watt	
	1960	1990
Filament lamps	12	12
Fluorescent lamps	75	100

Source: 'Efficiency of Electricity Use', *House of Lords Paper 37*, HMSO 1989; CIE Technical Committee TC 7-03.

6.3 Principal technologies

The following three changes in lighting practice, made possible by advances in technology, offer the greatest potential for energy saving whilst maintaining lighting standards:

* a change from filament lamps to compact fluorescent lamps;

* a change from mains-frequency fluorescent lighting to high frequency fluorescent lighting;

* the introduction of systems of need-control of lighting; that is, according to the need for light at a given place and time. This will usually be part of a BEMS (Building Energy Management system).

These three changes in lighting practice and their potential for energy saving are discussed in turn.

Figure 6.1 Improvement in lamp efficacy

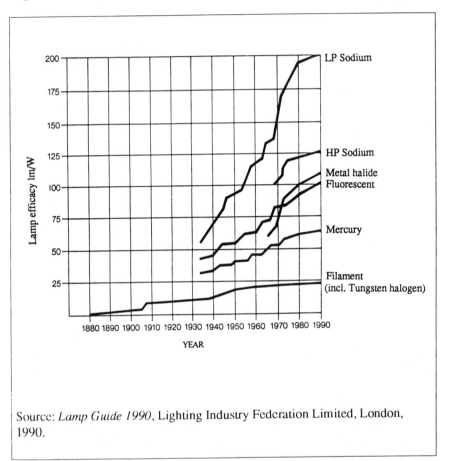

Source: *Lamp Guide 1990*, Lighting Industry Federation Limited, London, 1990.

There have also been continuing improvements in the efficiency of filament lamps, the development of the tungsten halogen lamp, and improvements in efficiency and colour rendering of discharge lamps, in particular the metal halide lamp and the high-pressure sodium lamp. But the greatest potential savings in energy are associated with fluorescent lamps, because they already account for 55% of the energy required for lighting in the developed countries, because of the potential for improvement of older fluorescent lamp installations, and because they can replace filament lamps, which account for a further 27% of lighting energy (Table 6.3 and Figure 6.1).

Figure 6.2 Compact fluorescent lamps

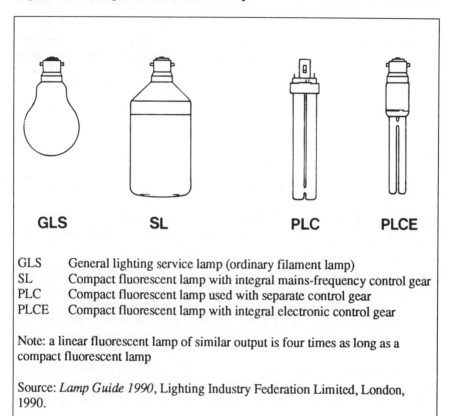

| GLS | SL | PLC | PLCE |

GLS   General lighting service lamp (ordinary filament lamp)
SL   Compact fluorescent lamp with integral mains-frequency control gear
PLC   Compact fluorescent lamp used with separate control gear
PLCE   Compact fluorescent lamp with integral electronic control gear

Note: a linear fluorescent lamp of similar output is four times as long as a compact fluorescent lamp

Source: *Lamp Guide 1990*, Lighting Industry Federation Limited, London, 1990.

6.4 Compact fluorescent lamps

The physics of the fluorescent lamp dictate that it should be long, typically over one metre. In the past, shorter lamps have had reduced efficiency and have been restricted to low power and low output. The research and development effort to find a successor to the filament lamp has produced solutions to some of the problems of earlier fluorescent lamp designs; continuing research is aimed at further improvements. Compact fluorescent lamps have folded arc tubes so that their overall dimensions are close to those of filament lamps. Examples are shown in Figure 6.2.

Table 6.4 Comparison of filament lamp with compact fluorescent lamps

	Filament lamp GLS 100W	Compact fluorescent lamps	
		SL 25	PLCE 20
Lamp cost (£)	4 (8 x 50p)	12	14
Electricity cost (£)	56	14	11
Total cost	£60	£26	£25
Energy consumed	800kWh	200kWh	160kWh

Notes: life of filament lamp 1,000 hours, compact fluorescent lamps 8,000 hours. Compact fluorescent lamp retail costs 1991/2, but likely to fall. Costs and energy are compared over a period of 8,000 hours, corresponding to one year continuous lighting or to four years with 2,000 hours lighting per year. Quantity of light about the same. Electricity cost 7p/kWh (typical 1992 UK domestic tariff).

In most industrial and commercial installations there will be an extra cost for lamp changing, which further favours the compact fluorescent lamp. For example, at £1 per point, the total cost is increased by £8 for the filament lamp, and by £1 for the compact fluorescent lamp.

Source: Philips Lighting UK.

Compact fluorescent lamps are about four to five times more energy efficient than filament lamps and last about eight times longer. This means that considerable savings can be achieved in both energy costs and maintenance costs, as illustrated in Table 6.4. Even if maintenance costs are not included, there will be a short payback period, provided that the lamp operating time per year is long enough. Eyre estimates that compact fluorescent lamps produce a rate of return of 38% for lamps run for 1,000 hours/year, corresponding to a payback period of about two years.[5] Commercial lighting with a typical operating time of 2,000 hours/year has a shorter payback period.

At first compact fluorescent lamps were proprietary, but there is an increasing move towards standardization, so that most lamp types are interchangeable. Lamp makers do not foresee a precipitate fall in the use of filament lamps, but are making a substantial capital investment in development and production of the new lamps.

Compact fluorescent lamps have ratings between 5W and 40W. They fall into two categories:

* Those with integral control gear and a conventional BC (bayonet) or ES (screw) cap. These lamps are suitable to plug straight into many luminaires which previously used filament lamps. The integral control gear was originally mains-frequency, but a trend is developing towards integral high frequency electronic control gear, with smaller lamp dimensions and reduced weight.

* Those which are designed for operation on separate control gear. One type of lamp has a two-contact cap with integral starter and is for mains-frequency operation. Another type has a four contact cap without starter: this allows more circuit options, such as operation on a high frequency ballast and in emergency lighting circuits.

Compact fluorescent lamps operate at a lower temperature than filament lamps, which makes possible new designs of luminaire using less material.

6.5 High frequency lighting

Fluorescent lamps, such as the linear type found in offices and some kitchens are usually operated at the frequency of the mains electricity supply, but their efficacy is increased if they are operated at higher frequency (Figure 6.3). For example, a lamp taking 50 Watts on high frequency has the same light output as a lamp requiring 58 Watts on mains-frequency.

The high frequency operation of fluorescent lamps has been made possible by recent developments in electronics. The conventional

---

5. N.J.Eyre, 'The Abatement of Gaseous Emissions by Energy Efficient Lighting', *Energy and Environmental Paper No. 2*, Energy Technology Support Unit, Harwell, March 1990.

Figure 6.3 Increase in lamp efficacy at high frequency

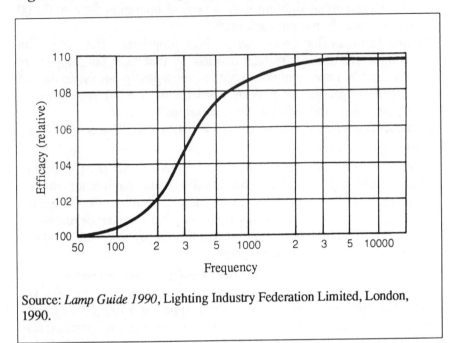

Source: *Lamp Guide 1990*, Lighting Industry Federation Limited, London, 1990.

mains-frequency ballast (50 or 60Hz) is replaced by an inverter circuit operating typically at 30kHz. Ballasts last much longer than individual lamps, and the extra cost of the high frequency ballast is usually recovered within the lifetime of the first lamp. For an example of costs see Table 6.5.

Conventional ballasts contain a kilogram or so of copper and iron. The change to electronics in high frequency ballasts means reduced extraction of raw materials, and reduced energy and pollution in their processing. At present discrete components such as transistors and resistors are used, but the next few years should see the introduction of integrated circuits which will make possible reductions in both size and cost.

Table 6.5 Comparison of conventional fluorescent lighting and high frequency lighting

	Conventional luminaire (2x58W 1500mm)	High frequency luminaire (2x50W 1500mm)
Luminaire power	140W	110W
Electricity cost	£98	£77
Energy consumed	1400kWh	1100kWh

Notes: comparison period 10,000 hours, a typical period for group replacement of linear fluorescent lamps. Quantity of light about the same. Electricity cost 7p/kWh (typical 1992 UK tariff, domestic and small commercial premises).

The cost of a high frequency luminaire is about £15 more than that of the corresponding conventional luminaire. The extra capital cost is usually recouped in the lifetime of the first lamp. The 20% saving in energy and energy cost is continuous. Where the conventional lamp is of a lower-efficiency type, the savings are greater. In addition, high frequency offers greater visual comfort, because of the absence of flicker.

Source: Philips Lighting UK.

6.6 Need-control of lighting

There are three principal opportunities for saving energy by need-control of lighting:

* integration of lighting with daylight ('Daylink');
* task-related regulation of lighting;
* presence-detector control of lighting.

In 'Daylink' the luminaires are regulated automatically to top up the daylight contribution. At any given moment, the electrical energy used is the minimum needed to bring the lighting to the required level. This is illustrated in Figure 6.4. With task-related regulation, the lighting level

Figure 6.4 Electricity saving with 'Daylink' system

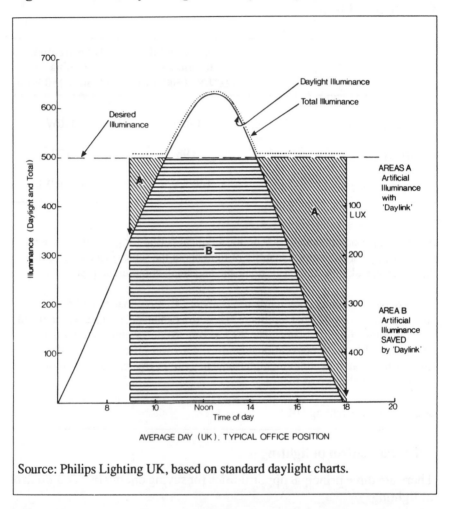

Source: Philips Lighting UK, based on standard daylight charts.

is adjusted according to the need of each area. For example, an open-plan office may be lit to 500 lux, but an area with filing cabinets could have the lighting turned down to, for example, 300 lux. It is also possible to cater for the different needs of individuals, an aspect to which increasing importance is being attached in the design of working areas.

Both 'Daylink' and task-related regulation have been made possible on a large scale by the recent development of high-frequency regulation lighting, an extension of high-frequency fluorescent lighting in which the electronic ballast adjusts the light output and input power. By regulating the light output and energy consumption of fluorescent lighting systems between 100% and a low level, high frequency regulation can provide savings in electricity of up to 30% in addition to those provided by high frequency operation, a total saving of about 50% as compared with conventional fluorescent lighting.

Presence-detectors are devices using, for example, infra-red radiation to detect whether people are present in a given area. When a room has been unoccupied for say 10 minutes, the detector gives a signal to switch off the lighting. The lighting is switched on again as soon as someone enters.

6.7 Principal market areas

Compact Fluorescent Lamps. In developed countries, compact fluorescent lamps are suitable for many applications in the domestic and commercial lighting sectors, including offices, shops, and hotels. For example, in the UK there are over 300 million domestic lighting points, of which less than one million are equipped with compact fluorescent lamps. In developing countries, where filament lamps are still relatively common in industry and in public lighting, compact fluorescent lamps can be used as replacements usually without the need for new luminaires; this should be an interim step until resources become available for a change to linear fluorescent lamps and to discharge lamps and for the development of the electricity supply system. In some countries there are wide fluctuations in supply voltage, and the move to electronic ballasts makes it easier to deal with this problem.

High Frequency Lighting. High frequency lighting, with linear fluorescent lamps, combined with need-control where appropriate, can be applied throughout almost all of the commercial sector, for example in offices and shops, hospitals and hotels. It also has applications in those parts of industry where fluorescent lighting is used. In the UK, there are over 100 million luminaires for linear fluorescent lamps, of which less than 1% are high frequency types. Most of the luminaires are more than ten years old, and suitable for updating.

There is potential for presence-detector and other control methods in almost all lighting sectors.

6.8 Pressures for increased uptake

Compact Fluorescent Lamps. There has been a steady growth in demand for compact fluorescent lamps within the commercial lighting sector, especially where the scope for energy saving is obvious, for example in hotel corridors with all-night lighting and in walkways in blocks of flats. Filament lamps are also being replaced with compact fluorescent lamps in areas such as shops, bars and restaurants, where the potential for substantially reduced costs is increasingly appreciated.

High Frequency Lighting. Where linear fluorescent lamps are used, and especially in office lighting, the advantages of high frequency lighting are beginning to be accepted by professionals, such as engineers and accountants. In addition to the energy saving, organizations see the value of the increased visual comfort (eg. absence of flicker), with its implications for reduced complaints about headache and eyestrain. Some organizations value the visual comfort even higher than the energy saving. At the top end of the luminaire market, therefore, the adoption of high frequency lighting is growing steadily.

Further savings in energy and costs are a major incentive to introduce high frequency regulation. But the concept is still new, and the demand at present is confined to the larger organizations in the private sector.

6.9 Constraints on uptake

Compact Fluorescent Lamps. The acceptance of compact fluorescent lamps in the very large domestic market is slow. A major obstacle is that it is difficult for the domestic user to balance energy savings against higher lamp costs. The initial cost to the domestic user is many times that of a filament lamp, but potential users should consider total costs including energy (Table 6.4). The lamp cost is recovered within about one-third of the lamp life. Some householders may appreciate that lifetime costs of compact fluorescent lamps are much lower, but may nevertheless be unwilling or unable to pay the initial cost of the lamp. For the biggest energy savings in lighting this is a difficulty that will have to be overcome, for example by electricity supply companies promoting lamps and spreading lamp costs over successive electricity bills. Such a scheme would be aided by the distributor's discount percentage, which represents a much higher sum for a compact fluorescent lamp than for a filament lamp.

As sales increase, the cost of compact fluorescent lamps will fall, though their greater complexity and material requirements mean than they can never be as cheap as filament lamps.

At present, integral compact fluorescent lamps are bigger and heavier than filament lamps, and may not be suitable for some luminaires designed for filament lamps. Also, some domestic users would prefer new lamps to look exactly like filament lamps. These problems will get steadily less with advances in lamp developments, and especially with the incorporation of the much smaller and lighter integrated circuit components.

Another unfamiliar characteristic is the warm-up period to full light output. This warm-up period has recently been reduced, and further developments should make it negligible.

In the UK the electricity supply companies are reluctant to take any action which will reduce domestic energy use and have been slow to promote compact fluorescent lamps. Some of them have drawn attention to the 'power factor', a characteristic which will be improved in the next generation of lamps.

High Frequency Lighting. Fluorescent lighting with linear lamps is dominated throughout the world by the low-cost batten[6] market, in which initial cost is the prime consideration, with energy savings often being overlooked. High frequency developments may not obtain a share above 10% in the batten market unless there is action by governments to favour the more energy efficient products.

High frequency regulation has the same problems of initial cost versus longer-term savings as has high frequency lighting. There are also the problems that extra wiring may have to be installed, and that some installers may be reluctant to handle something new.

6.10 Potential impact

Compact Fluorescent Lamps. The widespread use of compact fluorescent lamps could in theory cut the electricity used in domestic lighting by 75%, though a cut of 50% is a more realistic target. Even a 50% cut would require involvement by governments. Within the commercial sector in developed countries a saving of about 15% of the lighting energy is possible by replacing most of the filament lamps with compact fluorescent lamps, plus about 10% saving in the industrial

6. A batten is a basic lighting fitting for linear fluorescent lamps.

sector. As an example, for the UK the all-sector savings, based on the sectors of Table 6.2, would be about 7TWh a year.

High Frequency Lighting. The greatest percentage of electricity for lighting is in the commercial sector. The national energy saving in developed countries from high frequency lighting could be about 10% without official incentives, but there is a potential to increase the saving to about 25%, if suitable policy instruments were adopted by governments. Savings in the industrial sector could be about 20%. In the UK, for example, the total energy saving from high frequency lighting would be about 6TWh a year.

High frequency regulation could extend the national energy savings possible from high frequency lighting typically by a further 20%; in the UK from 6 to 7TWh a year. It would be an advantage if the two developments were run together, so that only high frequency regulated lighting is installed in future.

A simple change in practice such as the use of presence-detectors could reduce the national requirement for lighting energy by 5% or more.

In short, in the UK savings of 12-15TWh a year in lighting energy are quite attainable, given a national Energy Plan, without further advances in lighting technology. Without an Energy Plan, the savings would be less than 5TWh a year.

6.11 Policy mechanisms and options

Energy efficient lighting depends mainly on policies to support energy efficiency in general. An important development is the growth in many countries of taxes on extraction and on pollution. Progressive pricing of electricity, whereby the price per unit increases as the number of units increases, would also be an incentive by which governments could encourage the efficient use of electricity, and help low income households.

Building regulations should be updated to make energy efficient lighting mandatory, at least in utilitarian applications. Improved standards in new buildings will also influence practice in the refurbishment of existing buildings.

Governments should set an example by adopting a policy of replacing filament lamps with compact fluorescent lamps and installing high frequency lighting throughout their own offices and buildings. Wherever possible, high frequency regulation should be installed. In the UK alone there are over one million lighting points in government offices. Almost all use either filament lamps or conventional mains-frequency fluorescent lighting. A general directive is urgently required.

A further requirement is for regulations for retailers to display for domestic lamps not only the price but also the typical cost of electricity over a standard period (for example, 1,000 hours). This would demonstrate the savings to be made by a change from filament lamps to compact fluorescent lamps and would help to overcome customer resistance to the purchase price of the lamps.

Governments differ greatly in their approach. The UK government has taken only the minimal step of exhortation, via the Energy Efficiency Office. There are no regulations to support energy efficient lighting, there are no financial incentives/disincentives, and government buildings set a poor example.

A further problem which is particularly acute in the UK is that the privatized companies producing and retailing electricity are faced with conflicting objectives. On the one hand, they have a duty to their shareholders to maximize profits - which in the current regulatory system means increasing sales, but on the other hand they have a duty to humanity to promote more efficient use of energy. This is a dilemma which can only be solved by governments changing the regulatory framework.

In contrast, under schemes applied in parts of the USA and also in Canada, the Netherlands and Scandinavia, electricity supply authorities are given incentives to give away or subsidize compact fluorescent lamps, in order to reduce energy demand and requirements for new generating equipment. For example, in the USA the Southern California Edison Company and the New England Electric System have given away over one million compact fluorescent lamps to low-income homes, resulting in social benefits and reducing the need for new generating capacity. In Canada, Ontario Hydro provides subsidies to commercial and industrial users to change from filament lamps to compact fluorescent lamps. These initiatives occur in part because of different regulatory frameworks which promote 'demand-side management' (Chapter 14).

In developing countries also it could pay electricity supply authorities to lease or sell at low cost compact fluorescent lamps. As living standards rise, lighting is often the principal electrical load, and a load that peaks in the evening. The subsidy could be recouped in reduced generating capacity and in peak-load reduction.

In the USA the Federal Law for Appliance Energy Conservation includes requirements for the most common ratings of ballasts for fluorescent lamps. All new ballasts must have a high ballast efficacy factor. This law, together with federal funding for research, has given a strong impetus to the change to high frequency lighting.

Details of campaigns in several countries to promote energy efficient lighting, and comparisons of different forms of incentive, have been compiled[7] and should be studied for use in lagging countries, such as the UK. The most successful schemes are likely to be self-financing ones, eg. a levy on low efficiency installations which is repaid on a change to high efficiency.

Trade associations and professional institutions around the world are supporting energy efficient lighting. For example, in the UK the Lighting Industry Federation and the Chartered Institution of Building Services Engineers support promotional schemes such as the National Lighting Award. But only governments can introduce the legislation and financial incentives/disincentives that will make possible a new era in the efficient use of lighting energy.

New technology will require education through re-training schemes, for example for installers, which should be organized by employers and professional institutions. There should be a greater overlap of lighting courses, including those for architects and interior designers, with electrical and electronics courses. The institutions should put greater emphasis on continuing professional development.

6.12 Conclusions

The use of lighting will continue to grow as the world economy expands. Provided that the technologies described are adopted, this growth could mean no increase in the world demand for energy for lighting, with no reduction in standards of lighting.

Published estimates for national energy saving in lighting range from 20% to 90%.[8] In practice, the highest savings would depend on a steep

7. 'Proceedings of the European Conference on Energy Efficient Lighting', Swedish National Energy Administration, Stockholm, May 1991, and E.Mills, 'Evaluation of European Lighting Programmes', *Energy Policy*, April 1991.
8. For example, Eyre (op.cit.) estimates about 50% for the UK.

rise in the cost of energy. In developed countries, at current energy costs, a saving of about one-third in the energy for lighting is a practical proposition. For example, in the UK, the saving could be 12-15TWh per year, provided that there is government support.

Energy efficient lighting as a component in saving world energy stands out for three reasons: lighting is a significant user of energy and a high percentage can be saved; the technology already exists to produce these savings; and one-third saving can be achieved within a decade.

The three principal developments in energy efficient lighting are: the move from filament lamps to compact fluorescent lamps; the move from conventional fluorescent lighting to high frequency fluorescent lighting; and the move towards needs-control of lighting.

The worldwide response to energy and pollution problems must include both national and international plans to bring about energy efficient lighting. Well-framed regulations and financial incentives/disincentives, together with good examples from governments, would start a movement resulting in major energy savings; and lighting users would gain financial benefits with no reduction in lighting standards.

Building Energy Management

Nearly half of all energy consumed in industrialized countries is used in buildings. The total energy used in a building depends on the efficiency of the various enduses such as heating, lighting and air conditioning, on the building itself and on the behaviour of the occupants. The way these elements are dealt with is termed building energy management.

In addition to other individual options for using energy more efficiently (which include, in addition to technologies covered in this book, gas condensing boilers, heat pumps and small-scale combined heat and power - CHP), savings are available from improving the overall management and coordination of energy use in buildings, especially in larger buildings such as those common in the service and light industrial sectors. Building energy management systems (BEMS) exploit modern computing, control and sensor technologies to manage energy uses in buildings and their interactions more efficiently. Drawing on examples from the UK and European service sectors, this chapter illustrates the potential for BEMS and many of the issues involved in improving energy performance in these sectors.

There are substantial market opportunities for BEMS in Europe and elsewhere, with a large potential for energy saving without taking account of improvements in the efficiency of individual enduses. However, despite substantial progress uptake remains slow. Constraints on more rapid uptake include ignorance of the benefits of the technology, lack of industry standards, shortage of trained staff, lack of finance in some sectors, leasing structures and accounting practices.

Government initiatives in the UK have included the Energy Efficiency Demonstration Scheme, which has encouraged and demonstrated BEMS technology, and the setting up of the BEMS Centre. Government action is now concerned more with dissemination of information through the Best Practice Scheme of the Energy Efficiency Office; direct funding of projects has been reduced. Better energy management and increased takeup of BEMS could be promoted by a requirement for best practice or other minimum standards through the Building Regulations.

Buildings consume 30-50% of total energy demand in developed economies; in the UK buildings are estimated to account for nearly half (47%) of delivered energy, which equates to about 45% of primary energy inputs to the UK economy. Of this, about a third (13% of primary energy) is consumed in the service sector, which comprises the commercial sector, covering premises such as shops, offices, pubs and restaurants; and the public sector, which covers government, local authorities, education and health. Services, especially commercial premises, constitute one of the fastest growing areas of energy demand in developed economies.

Energy use in services has many features in common with that in light industry and even some domestic uses (such as housing estates), so that developments in the service sector also have wider implications. Energy is used for a wide range of purposes, such as space and water heating, lighting, air-conditioning, computing, printing, refrigeration, and in some cases electricity generation. Usually these components are used separately, with little attention given either to the energy use of the different components or of possibilities for merging and integrating the use of the different components. However, many buildings are often large enough to justify significant investment in systems and training to manage energy use and other building activities more effectively. Modern technologies enhance opportunities for doing so.

The treatment of buildings as integrated systems illustrates many of the general issues concerned with energy efficiency. Although many different technologies could contribute to building energy efficiency, their detailed consideration is outside the scope of this chapter. Consequently, following a brief review of some energy saving options, the chapter concentrates on building energy management systems.

In 1979 the Energy Efficiency Office identified the introduction of building energy management technology as a major opportunity for energy savings through the improved control of building services in non-domestic buildings. A portfolio of 44 projects was then initiated under the Energy Efficiency Demonstration Scheme (EEDS).[1] The objective was to assess the potential for building energy management. It was anticipated that of the total technical potential for savings of around £1,200 million a year, approximately a third could be saved by building energy management systems (BEMS) and improved controls, spread over these building sectors.

The results of the EEDS projects, which monitored the actual building energy consumption after the application of BEMS against set targets, show that the anticipated savings in these projects were considerably exceeded. In addition, several important pointers to achieving even greater penetration of the technology and to the avoidance of future pitfalls in its introduction to the building industry, were highlighted. The lessons learned form the basis of this chapter, but first a brief review of the composition of the service sector is in order.

7.1 The service sector

Compared to almost static national energy consumption from 1970-90, consumption in the UK service sector has grown rapidly, increasing by more than 20% since 1975; rapid growth continues. Energy consumption in buildings arises from demands for enduses such as space and water heating, lighting, catering, air-conditioning, refrigeration, and electrical equipment such as lifts, pumps, fans, computers and office machinery. The demand is normally expressed in terms of delivered energy, the amount delivered to the building boundary or meter.[2]

Within the service sector buildings are diverse in size, age and type. They range from 14th Century public houses to 1990s office blocks, from fast food take-aways to palaces, from crowded indoor swimming pools to empty warehouses. It is estimated that there are about 1.5 million

1. *Energy Efficiency Demonstration Scheme - A Review*, Report No: 1, Report No: 2, HMSO, 1984, 1988.
2. *Energy Use and Energy Efficiency in UK Commercial and Public Buildings up to the year 2000*, Energy Efficiency Series, HMSO 1989; L.Schipper, S.Meyers, and A.N.Ketoff, *Energy Use in the Service Sector: An International Perspective*, Lawrence Berkeley Laboratory University of California, LBC 19443, June 1987 (and several other later publications by Schipper).

Table 7.1 Estimated number and area of premises for 12 main sub-sectors in service sector (1985)

	Number		Area		Average
	1,000s	%	km^2	%	Area m^2
Commercial sector					
Office	220	14	65	9	300
Distribution	145	10	124	8	830
Shops	600	38	84	12	140
Catering	56	4	9	1	160
Pubs/clubs	94	6	21	3	220
Residential	43	3	38	6	880
Retail services	120	8	15	2	130
Other commercial	150	10	83	12	550
Sub total	1,428	93	439	53	
Public sector					
National govt/defence	18	1	49	7	2,700
Local authority	35	2	52	8	1,500
Education	42	3	105	15	2,500
Health	29	2	43	6	1,500
Sub total	124	8	249	36	
Total	1,555		689		430

Source: ETSU estimate based on *UK Energy Digest*, HMSO, 1986.

premises in the sector and Table 7.1 gives estimates of the number of premises and building area for the 12 main sub-sectors in 1985.

Over 90% of premises are in the commercial sector, although they account for less than two-thirds of building area. The largest sub-sector is shops, with nearly 40% of the total number but only 12% of the total area. The second most numerous is offices, with nearly 14% of the total number and 9% of area.

Although it has more than 30% of the total area the public sector accounts for only 8% of the number of premises. The largest-sized premises are in the national government and defence sector. Education has only 3% of premises but 15% of total area, reflecting the large average size of universities, colleges and secondary schools.

Large buildings, such as office blocks, department stores, hotels, cold-storage warehouses, hospitals and universities have proportionally higher energy consumption than other buildings. This is because they have more facilities, such as central heating systems and air-conditioning, restaurants, swimming pools, lifts and specialized equipment. The largest 5% of buildings are likely to use a substantial proportion of the total energy consumed.

Most buildings use more than one fuel. Electricity could provide all requirements, but cost reasons frequently dictate the use of several fuels.[3] Central heating and water heating are usually fuelled by gas, coal or oil. The most important enduse in the sector is space heating, which accounts for over 60% of delivered energy and over 40% of energy costs. Space heating therefore tends to receive most management effort in trying to reduce costs, both in existing buildings and in the design of new ones. Second in importance is lighting, at approximately 10% of total delivered energy but nearly 20% of energy costs. Water heating and cooking come third, each at about 8% of total cost (see Table 7.2).

7.2 Building energy use and options

The total energy used in a building depends on the behaviour of the people in it, on the building itself, and on the efficiency of the various enduses such as lighting, heating and air conditioning.

The importance of occupant behaviour and good basic management needs to be stressed. Better habits, such as remembering to switch the lights off and turning heating down instead of opening windows, could make a major contribution to energy saving. Better management, for example in the setting of heating controls and placing of refrigeration in cool areas (or at least away from heat sources), is important. Very basic technologies such as insulation, time switches for heating, and better lighting luminaires, are still not fully exploited.

Beyond these relatively simple steps, there are a number of particular technologies which can contribute to energy savings. The preceding chapters have discussed technologies for saving electricity in domestic appliances and lighting. Other technologies which could save energy

3. *The SCOPE 1991 Report - Study of Costs of Office Premises*, The Anderlyn Consultancy, London, 1991.

Table 7.2 Estimated expenditure on energy in service sector by enduse in UK, 1975-85, (1985 £millions)

| | Commercial sector | | | Public sector | | | Total Service sector | | |
	1975	1980	1985	1975	1980	1985	1975	1980	1985
Space heating	825	1,025	1,140	790	100	1,050	1,620	2,030	2,190
Water heating	140	160	175	170	210	220	310	370	400
Light	540	600	655	350	375	380	890	980	1,035
Cooking	155	160	190	120	160	160	270	320	350
Air conditioning	70	130	200	25	40	55	95	175	260
Other	210	360	545	85	145	200	305	500	740
Total	1,940	2,435	2,905	1,540	1,920	2,060	3,490	4,370	4,980

Source: ETSU estimate based on *UK Energy Digest*, HMSO, 1986.

particularly in service buildings include condensing boilers, heat pumps, and small-scale combined heat and power (micro-CHP).

Condensing boilers have additional heat exchange surfaces to recover the latent heat in the exhaust gases, which in conventional boilers is lost in the flue gases, and are consequently more efficient. They can be obtained in domestic, commercial and industrial sizes. Average annual efficiencies of condensing boilers are typically 80-85%, compared with 70-72% for modern conventional boilers which are of lower initial cost; they thus reduce gas consumption for heating by 10-15%. Based on 1990 UK gas prices, the payback period for new or replacement condensing boilers is 3-5 years, larger installations giving the shorter periods. For a large installation, such as in a hospital, the optimum choice may be a number of condensing boilers to operate at as near full load as possible, with conventional boilers used sequentially to optimize efficiency.

The uptake of condensing boilers varies widely between different countries according to energy prices, government and utility policies, and public awareness and attitudes. Uptake in the UK is small: customers are resistant to the significantly higher initial cost, despite the overall economic benefits, and reliable condensing boilers are difficult to obtain and maintain. Similar customer resistance is evident in other European countries. However, sales are rising rapidly in the US, encouraged by government regulations on energy efficiency.

Heat pumps are based on pumping cycles that extract heat from a low temperature source (the air or ground) and raise it to higher temperatures (the building). They are cost-effective in a few applications, such as dehumidification and heat recovery in swimming pools, and for space heating and cooling in a few commercial premises, such as offices and supermarkets. Several pilot schemes are undergoing trials which are seeking to control the loading of heat pumps via BEMS control. The duty of a heat pump can be reversed to provide building cooling; such heater-chillers are becoming increasingly common in air conditioning for buildings. The technical potential for energy savings from heat pumps is large in the service sector. The potential achievable in practice is difficult to estimate, but for heat pumps to gain widespread use for space heating and cooling, consumers' perceptions of their reliability and cost-effectiveness will need to be considerably enhanced.

Combined heat and power (CHP) is a technology which is already making a notable contribution to the electricity needs of the UK.[4] With some 500 CHP schemes already in operation, approximately 3% of Britain's electricity requirement is being met by CHP. The potential for environmental benefits through the reduction of carbon dioxide and other emissions are being increasingly recognized.

The cost-effectiveness of CHP is very sensitive to the ratio between electricity and gas prices, to the capital and maintenance costs of equipment, and to the profile and load factors of the electricity and heat requirements. Cost-effective applications are now being exploited in sites with continuous and high demand for electricity and for hot water, such as hotels, particularly those with more than 25 bedrooms,[5] hospitals and swimming pools, or to sites with very cheap gas, such as sewage works. A recent study, sponsored by the Department of Energy,

---

4. D.Green, *Combined Heat and Power and Electricity Generation in Industry*, Combined Heat & Power Association, London, 1991.
5. *Hotels & Catering Establishments in Great Britain*, HCITB, Longman, 1988.

showed that the economic or realizable potential for small-scale CHP is 320MWe in over 4,000 sites in the UK.[6]

Small-scale CHP units make use of a low cost fuel (usually gas) to generate both heat and electricity. They are slightly less efficient than boilers as heat generators but this disadvantage is more than offset by electricity savings. At the heart of a small-scale CHP unit is a reciprocating engine, modified to run on gas. These are relatively inexpensive, readily available and easy to maintain. Most have an electricity output in the range of 15-160kW, with a heat output up to three times the rated electricity output. However they have a relatively short life, perhaps only 20,000 hours of running time before a replacement engine is needed.

The UK government Environment White Paper[7] has singled out CHP as a specific area for targeted growth in the remainder of this decade - setting a goal of doubling CHP capacity in the UK by the year 2000 - with a further 2,000MW of CHP coming on stream.

If CHP and heat pumps are linked to the use of expert systems with BEMS, then this combination is likely to show the greatest advance in building control technology to the year 2000, with additional forecast electricity savings of 3-5% of current usage in the service sector.

7.3 Building energy management systems

Building services determine the quality of the environment experienced by building occupants. Managing the energy component of such services is known as building energy management, one aim of which is to reduce energy costs while maintaining services at the same level (or to improve the level of the services at the same cost), using a variety of energy efficiency measures.

Control is vital to energy management. The advent of microelectronics has greatly increased the ability to control services such as heating, air conditioning, ventilation, lighting, etc, although the information upon which control is based is frequently poor, through sensor inaccuracy, operator error or misuse. Energy savings of 30% or more can be achieved with improved control through BEMS. Apart from providing increased

6. *Market Projections for 1995 - Small Scale CHP in Buildings*, ETSU Reports Nos. GT/42/89 and GT/44/89, Environmental Technology Support Unit, Harwell, 1989.
7. Department of the Environment, *This Common Inheritance: Britain's Environmental Strategy*, HMSO,1990.

energy efficiency and reduced costs, BEMS provides an improved control of the working environment resulting in a happier workforce.

Control devices range from the simple 24-hour time switch, with two on/off settings a day controlling a small central heating system, to a microprocessor-based system for controlling a wide range of functions in an office building. Equipment is now available which can be used to control maximum demand for electricity, shed electrical loads, and to monitor and control heating and lighting by time of day, temperature and occupancy.[8]

Distributed intelligence systems are suitable for a single building or located at separate sites linked by a communications network to a central supervisor unit.[9] They allow for greatly increased flexibility, as data can be directly transferred between these outstations without the need for large amounts of processing power. Distributed intelligence systems are particularly suited to sites with multiple buildings or groups of geographically-dispersed buildings, sometimes in totally different parts of the country.

Distributed systems can initially be installed on a small scale, that is, with a few outstations, expansion of building and control functions being relatively simple. A central station can be added if required. Outstation costs can be as low as £350-£500, and overall costs can be shared between individual buildings.

The motive for undertaking energy efficiency measures is to reduce energy costs and obtain other improvements in equipment performance. A simple payback period is the most common investment criterion, and current energy prices will be an important factor. In practice, it is often difficult to relate cost-effectiveness directly to energy prices, as other factors may be involved. Refurbishment of shops, offices and high street premises may be undertaken to improve market image, to increase display areas, or to accommodate computerization, with improved energy efficiency as an additional benefit.

The use of BEMS alone could achieve an additional 15-20% saving in an already well-managed building. If lighting control systems are included, potential savings range from 30-70%. Although the full

8. P.Gardner, *Energy Management Systems in Buildings*, ETSU Report, Environmental Technology Support Unit, Harwell, 1984.
9. ibid; *Energy for Buildings*, IEA & OECD, HMSO, 1986. A review of microprocessor technology in buildings.

potential is unlikely to be realized, 65% has been achieved in practice. Realistically, energy savings of 20-40% can achieved, depending on the system and energy efficiency measures employed, giving pay back periods between two and two-and-a-half years. The addition of other technologies described above could increase this substantially.

The key feature of BEMS is the increased efficiency and energy savings they permit and the information they produce for management.[10] The distinction between management and control is important in this context. The ability to acquire, log and analyze data makes modern systems an indispensable tool for proper management of energy within the company or organization. This enables monitoring and targeting, which is the combination of systematic procedures for controlling energy consumption with a planned approach to the improvement of overall energy performance, resulting in more productive use of energy.

A BEMS has the additional advantage of pinpointing areas for more efficient management and maintenance functions to enable even more efficient energy use; the long-term benefits of historical data for use in the improvement of building performance should not be ignored. A further role for a BEMS is as an aid to more effective and efficient maintenance. Monitoring of plant condition and run times, for example, can be used to produce computerized planning of maintenance schedules.[11] On-condition maintenance which is carried out when condition monitoring dictates allows better use of maintenance resources.

7.4 The history of BEMS

Within the space of a decade in the UK, microprocessor-based control systems have progressed from being just a tool for the management of some of the services within a building to providing the effective control of all building services, including energy, in the whole estate. In 1981, the smallest system available would have probably cost tens of thousands of pounds sterling. Today, a system offering identical features can be purchased for less than £1,000. In fact the cost of processing has dropped to the point where it is reasonable to consider one controller per plant item, thus devolving intelligent control as far as it can go.

10. Gardner, op.cit.
11. BMCIS, *Energy Use in Buildings*, Building Maintenance Information Ltd, Kingston-upon-Thames, 1984.

The history and development of BEMS in different countries have been strongly influenced by both the prevailing political and economic climate which has led to distinct differences in design approach, particularly between the USA and European countries.

The USA can almost certainly claim to have operated the first BEMS in the early 1970s based on a centralized processor with all the executive control taking place at the 'centralized system'. These systems were based on mini-computer control stations and were most appropriate for installation in single large buildings such as the large corporate buildings so prevalent in the USA at that time.

The UK however possessed a different market with many smaller units operating within large distributed estates. The slightly later entry of the UK and European manufacturing companies into the market place presented the opportunity to incorporate the then emerging microprocessor into their products. Thus evolved the distributed intelligence system with the exploitation of the intelligent outstation.

7.5 European market size and structure

The Single European Market will bring about changes in the way the construction industry and contractors operate as they look to Europe for increased business and more suppliers that can provide support services in these countries. This will require, amongst other things, the ability to supply BEMS technology which is acceptable in the country concerned and which will therefore require conformity to certain standards.

Within the European Community, Germany and the UK together account for the majority of total sales and this is forecast to continue until 1993. An annual growth rate of 17.5% is predicted for this period but this will very much depend on the depth of the current recession in the construction industry.

The EC market is divided into three principal segments: the new construction sector having a 46% share, the refurbishment sector with 26.5%, and the retrofit sector with 27.5%.[12] Only Spain and Italy deviate from this pattern significantly. The commercial sector takes the largest share of business in all countries except Germany.

---

12. Refurbishment covers major refitting of an existing building, which could include rearrangement of office space, and new windows and building services. Retrofitting, which is on a smaller scale, is the replacement of one or more existing feature without complete refurbishment.

The reduced cost of the microprocessor in the last decade has opened up a wider market, which was worth in the UK around £100 million per annum in 1989 and growing at approximately 10-15% per annum. The two most developed markets, Germany and the UK, are virtually identical in the distribution of business by contract value and both have in the previous 2-3 years experienced rapid growth in small project business.

The exploitation of the domestic market in any country remains insignificant at the present time but considerable efforts are being made, notably in the USA, Japan and Europe, to provide controllers at a cost which will attract house owners and give a reasonable payback on their investment. Market research indicates that the needs of this sector are remarkably unsophisticated at this time and the primary requirement is for an integrated system incorporating fire and security. Development of intelligent systems to meet these requirements is progressing rapidly.

7.6 Implementation of BEMS

Energy management technology is of limited value in itself; it requires a close interaction with the people responsible for its operation if long-term savings are to be achieved. Once installed you cannot leave a BEMS to run itself and expect it to continue to produce savings indefinitely. In this sense it differs greatly from other building improvements such as roof and cavity insulation and double-glazing. A BEMS, including hardware and software, needs to be carefully specified to meet the complex needs of the services required.

The number of functional software options is very large, but it is important to select only those which can be economically justified. At present there are no performance standards for monitoring and control software and the performance of similar options offered by different suppliers is variable when judged against typical applications. Users require a scheme enabling them to distinguish between on the one hand reliable products obtained from those who have the resources and reputation to deliver a full service, from those that cannot.

BEMS are not a panacea for bad operation and practice. Other more cost-effective, less capital intensive measures may be appropriate before the BEMS can be justified. A thorough survey of target buildings and review of the building management practices will reveal if there are more cost effective measures and what the BEMS opportunities are. If the

BEMS option looks appropriate, it is essential to develop very clear objectives in relation to the type of application: what the system is intended to do, how it will be used and who will operate it.

Different types of application impose constraints on the optimum BEMS solution. The types of services included in the application will constrain the choice to systems with the appropriate range of applications software options, to suppliers with appropriate experience and possibly to systems approved by the appropriate authorities (eg. Fire Officer's Committee).

Annual budget limitations or a policy of gradual phased investment may restrict the choice to systems which are particularly flexible with respect to future expansion of capabilities, but which offer low-cost initial configurations. Recent initiatives towards compatibility between different manufacturers equipment are also relevant.

When a particular system is being considered, contact with existing users will reveal whether the supplier and his products have a successful track record. Have previous projects similar to the present one been successfully completed? Has the system fulfilled other user's expectation of costs, performance, ease of use and reliability?

Experience has indicated that several BEMS projects have been set up with quite unrealistic timescales for implementation. The user is understandably keen to start reaping benefits from the system and especially to make maximum savings during the next heating season. Equipment suppliers, anxious to please and impress, often agree to schedules which cannot be maintained in practice. Time slippage with its associated deterioration in relationships between customer and contractor, is then almost inevitable.

The prospective purchaser's existing plant and services often provide a potential source of difficulty. A BEMS needs to be interfaced with existing plant room equipment and during the course of installation defects that have lain hidden for years may be unearthed; for example mixing valves are often found to have exceeded their useful lives and in need of replacement.

Additional development work may be needed to meet particular requirements. This is likely to involve software rather than hardware and can result in considerable delay. The time and cost required both to develop and to prove new software is frequently underestimated and such requirements should be avoided wherever possible. An established, proven product may be a better alternative. BEMS and its associated equipment have many complex electrical and mechanical parts, several of which need proper regular maintenance if they are to function over long periods. Modern electronic components are renowned for their very high reliability, but faults can still occur.

Problems are most likely to arise with electromechanical equipment; disk drives and printers at the central stations and relays in the plant room are typical trouble spots. The simplest arrangement for maintenance is through a contract with the equipment supplier, usually based on annual payments of a fixed percentage of the initial capital cost.

7.7 Future developments in BEMS

The main future trend is likely to be a steady reduction in costs, as exploitation of microprocessor technology proceeds, as the benefits of increased production levels and lower costs begin to come through, and as there is fuller integration with other specific energy-saving technologies.

The continuing reduction in hardware costs will extend the cost effectiveness of the technology further down the buildings size range, but the limiting factor may be installation costs, which now account for up to 50% of total project costs. Some reductions however, are expected through greater exploitation of distributed processing and improved communication techniques, including mains-borne signalling (signals communicated through existing mains wires) which is finding more extensive use on most aspects of building services control, including heating, ventilation, air conditioning and lighting.

There are also likely to be substantial improvements in software with individual systems offering wider ranges of management functions, to increase their ease of use. Offline analytical programmes and computer spreadsheets will be more important in helping with the burden of interpreting data or proving monitoring and targeting procedures. Great scope exists for the development and use of expert systems that can make judgements and recommendations to the user and also provide assistance to less skilled operators of the systems to achieve the best solution to problems, eg. alarm situations.

There is also likely to be greater integration of energy management functions with other aspects of building management. Fire and security monitoring require similar data acquisition and control instruction transmission procedures as energy management, and are sure to be increasingly incorporated into future systems.

Development of the microprocessor-based hardware elements of BEMS technology has now reached the point where it is capable of satisfying all likely user requirements for the foreseeable future. Software development is generally not so advanced but it is progressing rapidly to match the capabilities of the hardware and management needs of future users.

The limited accuracy of existing sensors still limits achievable savings. Considerable scope exists for developing cheap, rugged and reliable linear temperature sensors for environmental control. Air conditioning control is beset with even more severe sensor problems stemming from the limitations of present humidity measurement techniques. More development is also required of components such as valves and actuators. Development of communications systems will not only benefit building energy management, but other electronic devices in the building such as facsimile transmission, word processors and conference videoing devices. Fibre optics are rapidly being adopted for the telecommunications industry; being light in weight, strong, extremely small and capable of transmitting vast quantities of information over hundreds of miles without amplification, make them an obvious choice for communications in the future. An alternative is radio transmission using packet radio linked through digital repeater stations and satellite links.

At the level of the individual sensor, technology for optical coupling is a key factor. It will be essential to develop the technology of micro optics, fibre fusion, fibre lapping and those using planar waveguide principles to achieve the full potential of this technology.

7.8 Constraints and policy options

The reasons for the increasing uptake of BEMS follow directly from the characteristics described above. They can reduce energy costs, improve the working environment, and provide management information to assist in the improvement of the overall performance of the building. Wider environmental benefits follow from the reduced energy consumption, but these rarely feature in the management decisions concerning BEMS, and in practice many other factors constrain the take up of BEMS.

As uptake increases with time, attitudes towards BEMS technology are likely to change gradually. It will be looked upon less as an immediate energy cost saving measure and more as a component of improved standards of management in the long term. By combining energy management with non-energy functions, a BEMS will become an essential aid to the effective utilization of all resources, including manpower and materials. As part of this, BEMS could reduce building energy consumption by an additional 20-40%, and perhaps even more if combined with better end-use technologies.

But the take-up of BEMS technology has been less than might be expected on simple economic grounds. Some of the reasons, identified in a report by the Parliamentary Select Committee on Energy, are:[13]

* lack of awareness: within each building category there are different levels of knowledge, effort and achievements;

* low priority given to energy conservation investments: in the public sector other activities are considered more important and in the commercial sector investment in productive capacity or turnover is considered crucial to survival;

* insufficient finance: local authorities, for example, would make further cost-effective investments but are constrained by shortage of funds;

* the present leasing structure presents a potential clash of interests (the landlord-tenant relationship) in that the benefits from the conservation investment may not accrue to the party which finances it;

* accountancy practices, which may require writing-off an investment over a shorter period than its physical lifetime, may militate against energy conservation.

Additional factors are:

* the shortage of appropriate skilled staff to exploit the opportunities available since BEMS technology crosses all the traditional boundaries of engineering. Education and training provides the key;

13. Fifth report from the Select Committee on Energy, Session 1981-82, 'Energy Conservation in Buildings', quoted in Energy Use and Energy Efficiency in UK Commercial and Public Buildings up to the Year 2000, *Energy Efficiency Series 6*, HMSO, 1989.

* the lack of industry standards to facilitate design and development of BEMS and thus to reduce costs.

Cost is not a major limiting factor, but budgetary constraints are limiting the uptake of BEMS in the public sector.[14] This could be overcome with the continued introduction of contract energy management, which is the provision of energy management services to a user under contract. Inability to finance potentially profitable investment directly could be dealt as part of contract energy management. Third parties, who may have access to funds at attractive rates, would finance the investment, which would be paid for either partially or completely out of revenue. Contract energy management is also a possible answer to the requirement for short payback periods. Another is the use of fiscal incentives such as direct subsidy or special tax concessions.

Government initiatives in the UK have included the Energy Efficiency Demonstration Scheme, which has encouraged and demonstrated BEMS technology, and the setting up of the BEMS Centre. Government action is now concerned more with dissemination through the Best Practice Scheme of the Energy Efficiency Office, in which direct funding of projects has been reduced. There is at present no government legislation requiring best practice in the technology. A minimum standard for building energy management could be set in the building regulations. This would be linked as appropriate with a requirement for mandatory standards and codes of practice.

To overcome the difficulty inherent in the landlord-tenant relationship, a mechanism needs to be found which will ensure that there is an equitable sharing of the costs and benefits of BEMS.

Although CHP is a specific area for targeted growth in this decade, it is unclear whether government believes this can be left to market forces. Since this is a desirable development for environmental reasons, mechanisms could be designed to encourage all technologies with the same advantages, including heat pumps and energy efficiency.

Policy mechanisms and options include better dissemination as is already being done to an extent through the Best Practice Scheme. Attitudes need to be changed of management as well as those of technologists, requiring re-education, which will not happen without government action. Contract energy management can solve a wide range

---

14. Audit Commission, *Saving Energy in Local Government Buildings*, HMSO, 1988.

of problems, but its extent will continue to be limited in the present market unless there are fiscal incentives. Again government action is required. Government needs to review its own attitudes to building energy management.

7.9 Conclusions

Best practice in building energy management offers very substantial energy savings. However, the building must be regarded as a system consisting of appliances and equipment under the control of a building energy management system. It must not be forgotten that it is the people within the building who need the services energy can provide. Education is an important factor, both of the users of the building and of the installers and operators of the equipment in it, including the BEMS.

BEMS are becoming increasingly common in service buildings, but there is still lack of appreciation of the benefits which can be gained from a systems approach to building energy management, which requires a carefully-tailored BEMS, the installation of energy efficient appliances and equipment to provide a comfortable environment for the occupants of the building.

PART III: SUPPLY-SIDE TECHNOLOGIES

Gas Turbine Systems

After the first successful gas turbine generated power in 1939, gas turbines developed rapidly. Initial applications were for military aircraft but in the early 1960s gas turbines were rapidly and successfully developed for generating electricity at times of national peak demand, for which their low capital cost and potential for rapid deployment at a time of low distillate fuel prices made them attractive. However, their relatively low efficiency precluded them from baseload generation, especially after the price rises of the distillate fuels they used.

More recent developments have shown that gas turbines can be integrated into more complex power cycles which have higher efficiency and lower overall lifetime costs than steam turbine plant. This, together with more rapid power station construction times and modular design, makes gas turbine plant an attractive alternative to conventional fossil fuel and nuclear plant.

Combined cycle gas turbine (CCGT) plants using natural gas at current prices are proven and highly competitive. With the development of gas resources and infrastructure and with the environmental benefits of gas, CCGT plants are expected to make a major impact in the energy industries over the next decade.

Although aircraft gas turbines are now very highly developed, application of aero technology to power production lags slightly behind, with opportunities for further improvement. Progress towards ever greater power, efficiency and economy continues, which will be to the benefit also of industrial plant.

Other fuels can be gasified to drive modern gas turbine cycles. Coal gasifiers yield high efficiency, producing less carbon dioxide and very low emissions of other gases compared to conventional coal plant, although such plant has yet to be proven economically. Biomass from sustainable sources is a possible fuel for gas turbines which would produce no net increase in atmospheric carbon dioxide. This plant would be particularly suitable for developing countries already rich in biomass.

No action has been required by government to encourage the installation of CCGT plant. However, government action to penalize polluting technologies would hasten development of CCGTs and the more advanced and complex gas turbine plant. Such action would be more environmentally effective and more cost-effective if it were international.

The concept of the gas turbine has been understood for a long time, but it was not until 1939 that the first industrial gas turbine was installed, along with the first flight of a jet aircraft in Germany, followed shortly by the UK. Following these successes, gas turbine technology developed rapidly, initially for military reasons. A subsequent powerful driving force was the need to develop economical, efficient and more powerful turbines for civil aircraft. Gas turbines driving propellers (turboprops) were developed for powering aircraft, but have been largely superseded by the greater power of jet engines (turbojets and turbofans).

Industrial and aircraft gas turbines developed separately until the 1960s, when a modified aircraft gas turbine was used to drive a power turbine connected to a generator, a pump or other driven unit.

8.1 Gas turbine technology

Gas turbines are internal combustion engines. Normally, they use either liquid petroleum fuels, or gaseous fuels such as gas from refineries, solid fuel gasifiers, land fill or natural gas. Although the gas turbine is an extremely complex piece of engineering, it is just a source of hot, high pressure gas, or gas generator. This can drive an aircraft forward. In an industrial gas turbine the hot, high pressure exhaust from the gas generator passes through a power turbine which is coupled to a generator,

Figure 8.1 Operation of a gas turbine

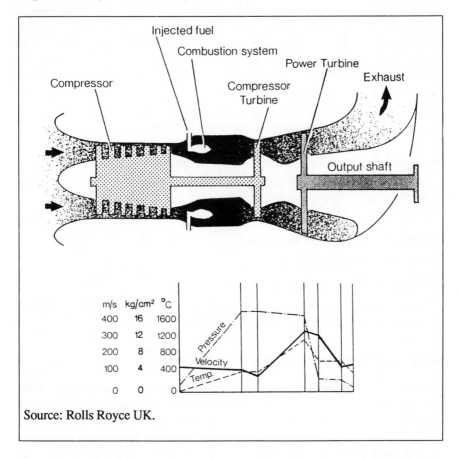

Source: Rolls Royce UK.

pump, or compressor. Gas turbines vary in size from under 100kW to over 200MW output power.

A typical simple cycle gas turbine (SCGT) operates basically as follows (see also Figure 8.1). In the gas generator, an axial flow compressor consisting of alternate rows of rotating and fixed blades draws in atmospheric air through an air intake and forces it through a convergent duct formed by the compressor shell into a casing where the compressed air is divided into separate flows for combustion and cooling purposes. A typically annular combustion chamber receives about 20% of the air flow for the combustion zone where fuel is injected and burned. Initial ignition is by electrical igniters, but these are switched off when the combustion becomes self-sustaining. The resultant expanding gas is cooled by the remainder of the air flow which enters the combustion chamber through slots and holes, and this reduces the temperature to an acceptable level for entry into the one or more axial stages of the turbine. The turbine drives the compressor to which it is directly coupled. Power turbine blading then converts the remaining high energy of the gas into net power at the shaft coupling to which the driven apparatus - an a.c. generator, oil pump, or gas compressor - is connected.

Gas turbines are by no means a new technology. Yet the late 1980s saw an explosive growth of interest in using gas turbines for power production, which is already being translated into market impacts. At least four factors have contributed to this:

  * improved system performance;

  * development of gas resources;

  * environmental pressures; and

  * changes in electricity markets.

The following sections review these factors, and examine the prospects for gas turbines in the 1990s and beyond.

8.2 Performance and efficiency improvements

Advances in the performance and efficiency of the gas generator are largely made in the aero engine field, where the intense competition, and massive market, ensures an investment in R&D which is probably unparalleled elsewhere. This drive for output, efficiency and reliability is matched only by the drive to achieve these improvements with minimum increase in cost.

Thermodynamic efficiency is a function of the maximum cycle or firing temperature and of the temperature of the exhaust. Both improved turbine materials and improved turbine cooling technologies have been very significant in advancing gas turbine performance by increasing the maximum firing (or allowable) fuel combustion temperature. The potential for reductions in exhaust temperature to increase efficiency are limited in simple open cycle gas turbine plant; the largest potential improvements are gained through the use of combined cycles and combined heat and power, as discussed below, but improvements in the gas turbines themselves remain important.

Major performance and efficiency improvements are not limited to increased turbine operating temperatures. Remarkable advances also continue to be made in component efficiencies. Improvements in compressors and turbines, and the reduction of other internal losses, can have a powerful effect on overall machine performance. Gas turbines in the late 1960s frequently had efficiencies as low as 20-25%;[1] subsequent developments have increased efficiencies of these open-cycle systems, and figures over 35% are common today. Industrial turbines to exceed 40% generating set efficiency are being developed by a number of companies.

Costs, reliability and ease of maintenance all have to be weighed along with efficiency, and with these improvements in performance, corresponding advances have been made in operating lives and condition monitoring of gas turbines to assess the need for overhaul or repair, boosting their availability and reliability still further. Aero-derived industrial gas turbines have achieved availabilities in excess of 97% in service over many years.

Gas turbines are flexible machines which can lie at the heart of more complex systems that provide greater overall performance or other different characteristics than that of the gas turbine alone. This is illustrated schematically in Figure 8.2, which shows the gas turbine as the central component to a range of options both in the fuel input and the use of output gases. Alternative fuel inputs are discussed towards the end of this chapter, as most current interest centres on the use of natural gas (or distillate fuels) which can be injected directly into the combustion chamber.

In a conventional gas turbine system much of the energy is exhausted to atmosphere in the form of heat and thus wasted. However, the greatest improvements in overall efficiency have been achieved by adding a waste heat recovery boiler in the exhaust system. This allows the

---

1. All generating efficiencies in this chapter are quoted on the basis of net heating values (lower calorific values). Gross heating values (upper calorific values) would yield figures about 10% lower for natural gas fuels (see Grubb, *Energy Policies and the Greenhouse Effect, Volume One: Policy Appraisal*, Dartmouth, Aldershot, p.434).

Figure 8.2 Alternative gas turbine systems

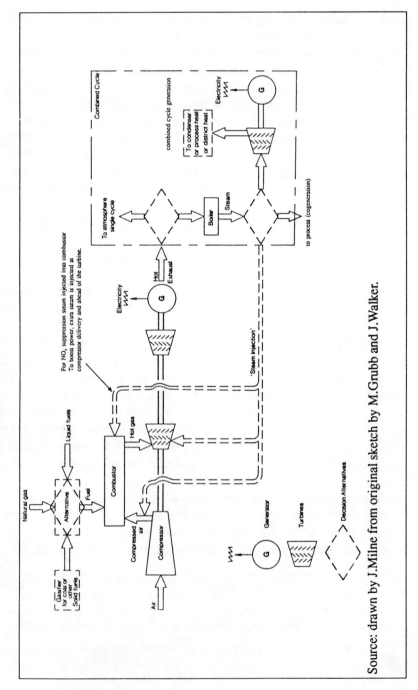

Source: drawn by J.Milne from original sketch by M.Grubb and J.Walker.

previously wasted heat to be used to generate steam either for industrial processes ('cogeneration')[2] or for boosting the overall efficiency of power generation. By making use of the exhaust heat, cogeneration systems can exploit about 90% of the fuel input as useful energy.

An alternative way of boosting power generation efficiency is by directing the steam raised by exhaust heat to a steam turbine generator. The combination of gas turbine and steam cycles gives the name of 'combined cycle' gas turbine (CCGT) system. The most recent CCGTs can achieve electrical efficiencies in excess of 50%. Although the addition of a steam turbine and other associated equipment can as much as double the capital cost as compared with a simple cycle gas turbine, such plant are still much cheaper than conventional power stations and the overall generating costs are highly competitive, as discussed below.

An alternative option is steam injected systems which inject the steam raised in the exhaust waste heat recovery boiler back into the gas turbine, which can boost the power and efficiency of the gas turbine itself. This technology was first introduced as a means of reducing the emissions of nitrogen oxides (NO_x) from the gas turbine for which the mass flow of water required is about equal to the fuel mass flow. But injecting steam in much larger quantities - a seven fold increase - can increase output by up to 50% without the additional capital expense of a steam turbine. Efficiency also improves significantly although not to the levels which can be realized with combined cycle.

The leap in power and efficiency arising from steam injection has its cost. Water for gas turbine steam must be extensively treated, necessitating expensive treatment plant and the steam is totally lost to the exhaust. In addition maintenance costs are increased although by precisely how much is currently difficult to assess since the technology is still in its infancy. This and other technologies may be attractive to less developed countries due to the lower capital requirements compared with combined cycle or steam turbine plant. However, availability of water, particularly to boost power, may be a problem in many areas, since a single 50MW plant would require about 70 tonnes of water per hour which in addition must be treated to a high standard of purity. Furthermore, design modifications are required to the turbine.

---

2. Cogeneration is the process also known as 'combined heat and power' (CHP) referred to in the demand-side chapters in Part II.

Combined cycles also require large quantities of cooling water, but river or sea water would be adequate, as for conventional steam plant. Steam-injected systems are commercially available and have been installed especially for cogeneration applications in the US, but for electricity generation the lower capital cost is insufficient to compensate for their substantially lower efficiency (typically about 43%).

However, system performance can be improved by cooling the air during the compression process. This 'intercooling' reduces the amount of work required for compression and hence the net output is increased. When combined with steam injection, efficiencies comparable to those of combined cycle plant are attainable at significantly lower capital cost.[3]

Complex systems such as this have yet to be developed commercially, and the advantages over combined cycle plant may be too marginal for manufacturers to put up the development capital which would be required. It is combined cycle systems fired by natural gas in which gas turbines and steam turbines are used jointly in a high efficiency system which are expected to make the most dramatic impact on the energy industries in the next decade, and the role of alternative gas turbine systems beyond this remains uncertain.

8.3 CCGT costs and economics

The investment costs of gas turbines are much lower than conventional power generation, even in combined cycle. This saving doesn't just occur in equipment cost, it is also a question of the acreage of ground required, the reduced installation costs and lower requirement for site work, and of course the lower cost of servicing finance resulting from the shorter building programme.

Figure 8.3 shows an estimate of the different contributions towards the overall costs of energy from CCGT, and compares them against various other conventional generating options. The higher fuel costs as compared with coal and nuclear are much more than offset by the lower investment and operation and maintenance costs.

3. Detailed discussion of alternative gas turbine cycles is given in Robert H.Williams and Eric D.Larson, 'Expanding Roles for Gas Turbines in Power Generation', in T.B.Johansson, B.Bodlun, and R.H.Williams (eds), *Electricity: Efficient End-Use and New Generation Technology, and their Planning Implications,* Lund University Press, Lund, Sweden, 1989.

Figure 8.3 Typical cost breakdown for CCGT and conventional sources

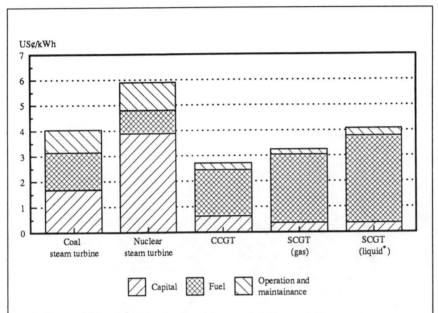

Note: *liquid fuelled SCGT will cost more to generate electricity and the efficiency is slightly lower with liquid fuel. Where gas infrastructure already exists it is the cheaper fuel option, however, due to the cost of new pipe for peaking plant, new systems will opt to burn liquid.

Assumptions: 30 year plant life; gas price US$2.5MBtu (a middle of the road US generator fuel price. US prices are very low at the time of writing; UK gas prices are 50% higher); 47% efficient CCGT plant, 32% efficient (typical) simple cycle gas turbine (SCGT) plant; 7,000 hours running per annum; no taxes included; nuclear fuel price US$0.84/GJ, nuclear plant efficiency 33.6%; coal steam plant efficiency 35%; coal price US$1.50MBtu, diesel price US$3/MBtu.

Source: Rolls Royce plc, UK (unpublished).

Note also that the simple cycle gas turbines are substantially cheaper to install, so that low-load uses (such as for peaking duty) favour simple cycle gas turbines as capital cost is more significant as operational time

Figure 8.4 Fossil fuel and biomass generation costs, and impact of carbon tax

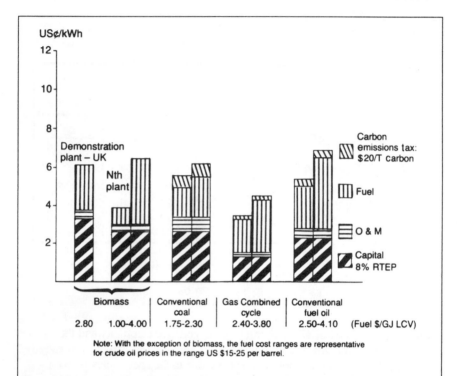

Note: With the exception of biomass, the fuel cost ranges are representative for crude oil prices in the range US $15-25 per barrel.

Note: Nth plant is demonstration plant and 'learning effect'; ie. mature production amortizing tooling costs. RTEP is Real Terms Earning Power, equivalent to the real interest rate. LCV is net heating value.

Source: P.Elliott, R.Booth, 'Sustainable Biomass Energy', Selected Papers, Shell International Petroleum Company, Ltd, London, December 1990.

is reduced. Indeed, one important advantage of gas turbine installations is that they can readily be expanded over time: gas turbines can be installed with the option to add waste heat recovery boilers, steam turbines and additional alternators at a later stage if electricity demand trends suggest that the incremental investment is viable.

Obviously the detailed figures vary with conditions and assumptions and above all with gas prices; Figure 8.4 shows an alternative estimate, which compares the costs against alternatives for a range of fuel prices (and illustrates the possible impact of a small carbon tax). But the broad conclusions are widely accepted and robust at the gas prices current at the beginning of the 1990s, typically around US$2/MBtu[4] as delivered to generators in the US, and 50% higher in most of Europe. The International Energy Agency summarized the current economic standing of CCGTs as:[5]

> The economics of gas in electricity generation is thus dependent on the discount rate, the price of gas and uncertainty about future demand ... gas is very competitive at a 10% discount rate because of its low capital costs: it is competitive with coal in low coal price regions, up to a gas price of $3.50/MBtu, and with other coal and nuclear options at gas prices between $4 and $5 per MBtu ... for some utilities gas is not economic at prices above $3/Mbtu because of lower discount rates and certainty about the future; for others gas could be economic above $4.50/Mbtu.

8.4 The availability of natural gas

The second major stimulus to the gas turbine market is the increasing availability of natural gas and its improving price against other fuels. Whilst gas turbines are by no means restricted to using natural gas fuel, gas has significant environmental advantages and the economic attractions already noted. Most of the current interest surrounding gas turbines is based on the use of natural gas.

There is a far more general acceptance now that natural gas is not scarce and the increasing level of resource availability discussed in Chapter 2 has led to a more relaxed view about its use for power generation. Of course it is not only a question of the reserves, it is also a matter of improved transmission and distribution systems bringing access to gas far more widely and with an increasing confidence of security of supply as the gas companies spread their contracts around various supply sources.

---

4. 1 million British thermal units = 10 therms = 1.05PJ. $1/MBtu = $5.8 per barrel of crude oil.
5. International Energy Agency, *Natural Gas: Prospects and Policies*, IEA/OECD, Paris, 1991.

These changes in the gas market are set to continue, with European Community acceptance of the use of gas in power generation, and moves towards third party access and integrated pipeline networks which may liberalize gas transmission. The moves may also ease further the negotiation of supply contracts, with an increasing move away from prices fixed relative to competitive fuels, towards competition between different gas suppliers, otherwise known as gas-to-gas competition.

As gas usage increases prices may rise, reflecting a tighter market and the additional costs of accessing new resources. Short-term fluctuations, reflecting supply/demand imbalances arising from constraints on existing delivery infrastructure, are likely; for example, in 1991 British Gas imposed a sudden 35% rise in industrial gas prices, sufficient to deter the rush towards CCGTs, because of fears about being unable to meet demand.

Concerning longer-term costs, Stern[6] cautions that 'the costs associated with current future gas developments are both difficult to identify and impossible to generalize'. Existing supplies to Europe amount to about 70BCM a year, which could support considerable expansion; but Stern suggests that in Europe new projects 'which would add [further] supply increments of 20-40BCM per year will probably require a delivered price greater than $4.00/MBtu'.[7] The economics of CCGTs for baseload supply at this point might become marginal, and sensitive to financing conditions, relative to other options such as the possible advanced coal technologies discussed in the next chapter. Equally, attempts to secure finance for projects necessitate long-term fuel contracts and these may prove difficult to settle if future fuel prices become less certain. In this context, one of the great potential advantages of the gas turbine is its fuel flexibility, as discussed below.

8.5 Environmental issues

Growing worldwide consciousness of environmental issues and indeed the political championing of the environmental cause is now also bringing a major encouragement to the use of gas-fuelled power generation systems, with even partial gas operation of existing coal-fuelled thermal stations in the USA under investigation as a way of

6. Jonathan P.Stern, *European Gas Markets - Challenge and Opportunity in the 1990s*, Dartmouth, Aldershot, 1990, p.68.
7. ibid.

Table 8.1 Gas turbine emission figures (g/kWh)

	NO_x	CO
Single-cycle gas turbine (SCGT)	4	0.3
SCGT (with steam injection)	0.5	1.5
Combined-cycle gas turbine (CCGT)	2.8	0.2

Emissions vary dramatically with engine type. These figures are intended to be a typical rather than representative of a particular engine in order to give a guide. Note that fluid injection reduces NO_x levels, but tends to increase CO levels.

Source: Rolls Royce, UK, unpublished material.

reducing peak pollution levels. The concern to reduce or eliminate sulphur dioxide and other emissions is putting up the cost of conventional coal plant and reducing its efficiency at a time when the costs of gas turbine and combined cycle power plant are reducing and efficiencies rising. See Table 8.1 for a breakdown of emissions with different gas turbine technologies.

→ The reduction of nitrous oxide emissions from gas turbines by the use of demineralized water injection into combustion systems, or by catalytic reducers in the exhaust, or both, is expensive. However, it is expected that NO_x levels attainable by these techniques will be equalled or bettered by improved combustor technology, thus making water injection or catalytic reducers unnecessary.

→ The concern for the greenhouse effect is also supporting the use of natural gas since it produces about half as much carbon dioxide per unit of electricity delivered as coal-based generation because of the lower carbon content in gas and the higher conversion efficiency.

In the foreseeable future there will be no commercially viable technology capable of removing CO_2 from power plant exhausts, so the most effective way of reducing CO_2 emissions is to improve the efficiency of fuel consumption.[8] Therefore power generation using high efficiency combined cycle systems and cogeneration can be expected to increase in significance over the next decade, particularly using the more environmentally benign natural gas as the primary fuel.

Another welcome environmental aspect of gas-fuelled power generation, of course, is the elimination of coal delivery and handling systems and their replacement by an underground piped fuel supply.

8.6 Electricity market developments

Another factor which may prove to have considerable significance for gas turbine demand is the way in which electricity markets are developing. In developing countries and eastern Europe, chronic shortages of capital will favour gas turbine plant where adequate gas is available and the conditions can be arranged for transfers of the relevant technologies and expertise; increasing concerns about the environmental implications of global energy developments based upon traditional energy technologies may increase the scope for such technology transfer arrangements. But the greatest current interest is in developed countries.

In developed countries, the driving financial force is not so much that of capital shortage as the implications of trends in electricity markets combined with utility reforms. Electricity demand growth continues but at a slower pace, and uncertainties about underlying growth trends, the possible impact of energy efficiency policies, and the rate at which old plant will have to be retired, all add to uncertainties about the scale and timing of requirements for new generating plant. Gas turbines and CCGTs are particularly well placed to meet such demand because even a large gas turbine combined cycle station can typically be producing power within two-and-a-half years from date of order, and as little as 20 months may be possible. The shorter lead times of gas turbine plant allow an accumulation of generating capacity far better matched to the changes in demand.

---

8. If oxygen is used instead of air in coal gasification, a shift reaction using steam can transfer nearly all the energy in the coal to hydrogen and as concentrated stream of CO_2. This can be absorbed into a concentrated liquid which can be pumped away. This process is potentially far cheaper than the removal of CO_2 from flue gases. Initial estimates suggest that only 10-20% may be added to the costs. However, disposal of the CO_2 stream may present a problem. (M.Grubb et al, *Energy Policies and the Greenhouse Effect, Volume II: Country Studies and Technical Options*, Dartmouth, Aldershot, 1991, pp.99-101).

The implications of utility reform, as outlined in Chapter 2, are perhaps even more important. The increasing role of private finance in utilities radically changes attitudes towards power investments. Private generators are wary of high-cost, long lead time plant, and are attracted by the low capital and marginal costs and low risk solution offered by combined cycle plant which enables them to achieve a more than acceptable return on investment.

The impacts have been most clearly demonstrated in the UK, where the government in 1988 published a White Paper outlining its plans for privatizing and liberalizing the electricity industry,[9] a process which was largely completed by late 1991. In 1988, the Central Electricity Generating Board was still planning to meet projected demand growth with a combination of conventional coal-fired power stations and nuclear power (having almost completed a two-year planning enquiry for a new nuclear station). CCGTs were entirely absent from utility plans. Three years later, with the industry privatized, all plans for new coal and nuclear stations had been dropped. In their place (excepting proposals for small capacities of renewable sources), by autumn 1991 grid contracts had been signed for 16,000MW of CCGT plants, of which over 6,000MW were already under construction, with a further 3,000-6,000MW firmly planned and likely to be financed. Whilst it is unlikely that any other electricity supply systems will go through such a dramatic transition, the direction of the trend seems inescapable.

8.7 The market for gas turbine plant

The demand for new power plant is likely to grow in almost all regions during the 1990s, for the reasons discussed in Chapter 2. Demand growth continues in the most developed regions of the world, where substantial existing capacity is also due for retirement; eastern Europe and the former Soviet Union may become chronically short of power (partly as a result of nuclear power station closures); developing countries continue to be chronically short of power, as well as capital. The need for new baseload generating plant is likely to total hundreds of thousands of megawatts over the decade.

Quite apart from environmental concerns about nuclear power generation and the decommissioning costs, the long construction times in this industry mean that any nuclear stations that will be operating in

9. House of Commons White Paper *Privatising Electricity* Cm 322; HMSO 1988.

the year 2000 must be in the planning, construction or operating phase now. Therefore any increase in capacity requirements over the next decade (over and above that already ordered) will have to be met mostly by gas-fired or coal-fired generation. Gas-fired generation is not only economically more attractive in regions with access to adequate gas; traditional coal plant is increasingly problematic especially in the more environmentally sensitive societies and, as discussed in the following chapter, adequate 'clean coal' technologies are insufficiently developed to take their place as yet.

The four factors discussed in this chapter - changing market requirements in terms of increasing environmentalism and industrial structural changes, the increasing availability of natural gas, and technological advances in the field of gas turbines combine to suggest that gas turbine technologies will take a substantial share of baseload generating capacity globally during the 1990s and beyond. One estimate is that the USA and Asia will contribute about 30% each of the gas turbine demand, with Europe accounting for about 20%.[10]

In addition, gas turbine systems have significant advantages for a range of more specialized applications. They are not only easy to buy and install, they also demonstrate operational flexibility, with start up and shutdown times significantly better than large conventional thermal plant, and with well proven reliability and the high availability that is associated with modular based machinery. Their smaller size also makes them particularly suited to systems making use of otherwise wasted exhaust heat.

Utilities and entrepreneurs will add efficient combined cycle systems based on 100MW+ gas turbines for base load capacity, whilst industry will exploit the capability of smaller gas turbines (50MW and below) to meet high site electricity and heat requirements. Cogeneration systems will improve energy efficiency and are likely to be encouraged by governments for environmental reasons.

Expanding cogeneration will reduce the requirement for utility base load capacity and increase the requirement for utility peak load capacity. This is because cogenerators will provide their own power for much of the time, but will require access to guaranteed standby capacity for their sites when equipment is being maintained or suffers a failure and cogeneration equipment power output may have to be 'topped up' by the

---

10. Source: unpublished material, Rolls-Royce plc, UK.

grid at times of high site demand. Thus the utility no longer supplies the steady base load demand, but is required to hold reserve capacity to meet sometimes irregular peaks. This may precipitate orders for simple-cycle gas turbines by utilities as they are cheap to install and therefore a capacity margin can be maintained without tying up large amounts of capital. Aero-derivative gas turbines are also able to achieve maximum load in under 5 minutes from cold and thus are ideal for peak lopping purposes. Of course, penal rates for emergency power demand may discourage cogeneration and it remains to be seen how this problem is handled by the industry.

The increasing gas demand over the next decade will necessitate increased production and transmission facilities and the gas turbine will also benefit from this growth. Aero-derivative gas turbines in the 10-30MW power range will enjoy increased business on compressor stations on natural gas pipelines and for offshore oil and gas production.

Improving efficiency and the introduction in the early 1990s of low emissions combustors will ensure reduced environmental impact in these activities without recourse to selective catalytic reduction or water injection which may be impractical for compressor stations and offshore platforms.

8.8 Alternative fuels

A final potential advantage of gas turbine systems is likely to gain importance in regions without access to adequate cheap natural gas, where efforts are made to keep fuels diversified, or if gas prices rise substantially. As illustrated in Figure 8.2, in principle a range of different fuels can be used to drive gas turbines.

Most of the energy in coal can be extracted either as a high pressure gas stream which can directly drive a gas turbine (from pressurized fluidized beds) or as a vaporized fuel which can be fed into the combustion chamber. The coal gasification side of the plant is expensive, meaning that it may only be possible to justify a plant of this kind on a fairly large scale, although this is by no means certain. The Coolwater plant operated by Southern California Edison has shown the technical viability of the IGCC (integrated gasification combined cycle) system and the recent Dutch decision to build a 250MW demonstration plant is intended to evaluate its commercial viability. Such technologies are discussed in the next chapter.

→ Another fuel source likely to become attractive in the next century is biomass. A range of different possible biomass fuels can be considered, from existing agricultural waste products (for example, from the sugar cane and pulp and paper industries) through to dedicated forest and herbaceous crops. Biomass is easier to gasify than coal and has a lower sulphur content, and (unless it is taken from primary forests) it does not contribute to atmospheric CO_2 build up since the CO_2 produced by the combustion process is effectively taken in by the growing plant crop.

Because biomass is a diffuse source it is more economic to consider smaller-scale power plants (up to about 50MW) than full scale combined cycle systems. On these grounds, various authors[11] have argued forcefully for a commitment to develop a commercial 'BIG/ISTIG' ('Biomass integrated gasifier / intercooled steam-injected gas turbine' system) which could be particularly (though not exclusively) appropriate for meeting power needs in developing countries. Figure 8.4 above illustrates an estimate of the economics of biomass-based systems compared with fossil fuel stations for a range of biomass prices.[12]

→ The option of an advanced cycle gas turbine burning gasified biomass has received considerable attention, at least in part because of its environmental attractions. One estimate suggests that biomass-fired plant may in the long term provide for at least 10% and possibly as much as 30% of world electricity.[13] However, in the absence of appropriate environmental taxes and further development efforts, such systems will not be commercially viable until the prices of conventional fuels rise considerably except in certain circumstances.

8.9 Policy issues

The key policy choices which affect gas turbine development are:

  * a utility regulatory environment which is favourable to environmentally cleaner technologies with low risks, low capital costs and short lead times;

11. E.D.Larson, P.Svenningsson, I.Bjerle, 'Biomass Gasification for Gas Turbine Power Generation', in Johannson et al (eds), *Electricity*, op.cit.

12. For further discussion of biomass technologies and prices see M.Grubb et al, Vol.II, op.cit., pp.103-113.

13. P.Elliott, R.Booth, 'Sustainable Biomass Energy', Selected Papers, Shell International Petroleum Company, Ltd, London, December 1990.

* development of gas resources and infrastructure;

* conditions which favour environmentally cleaner technologies, including emissions legislation, particularly on sulphur levels;

* energy and particularly carbon taxes covering electricity production.

In many developed countries, the first three of these has already occurred, which is why the use of gas for power generation will grow rapidly. As yet, few of these changes have taken place in developing countries, and consequently the growth of CCGTs will be slower there, despite the lower capital cost.

Carbon taxes would accelerate the introduction of CCGTs. Figure 8.4 above compares the different cost components of gas turbine cycles with natural gas and biomass against conventional sources, for a range of fuel prices, and illustrates the impact of a relatively small carbon tax of US$20/tC. This further improves the relative economics of CCGTs, and also of biomass to a greater degree.

However, it is unlikely that such changes alone will be sufficient to encourage other possible gas turbine developments (for example, ISTIG, coal gasification, biomass) to the market, because the benefits over CCGTs may not be enough to make companies confident in incurring the development costs and risks. Some level of government support for development and demonstration may be required. Alternatively, or in addition, credits for absorbing CO_2 may be sufficient to stimulate biomass developments based around gas turbines in some regions.

8.10 Conclusions

The gas turbine is likely to fare extremely well over the short-to-medium term as the world grapples with environmental problems and growing energy demands. To put the potential role of the gas turbine in perspective, the installed generating capacity of the UK is approximately 60,000MW; current annual global orders for gas turbines for power generation are running at around 25,000MW and this is likely to grow considerably.

The reasons are at present mostly to do with the rapid improvements in gas turbine systems themselves (especially combined cycle plant); the ready availability of natural gas; its environmental advantages; and changes in electricity markets which further enhance the advantages arising from the lower cost of generation from combined cycle plant combined with the speed with which gas turbine plant can be planned, built and commissioned by comparison with conventional coal, oil and nuclear-powered plant. The drive for higher efficiency, particularly of the gas generator, will continue.

A range of further developments is possible, including more advanced systems which avoid the steam turbine cycle, and applications to alternative fuels. None of these developments beyond existing and proven CCGTs will take place without positive action by government and international agencies. The pace of development and the role of fuels other than natural gas and technologies other than CCGTs will depend upon the nature of incentives or support from government. But in many areas, even without such support or further policy changes, gas turbines seem set to play a major role in future electricity markets, as a central generating option for a range of applications.

Clean Coal

Coal is and will remain a major fuel worldwide. But the environmental impact of burning it will ensure ever stiffer controls. In conventional plant for electricity generation, this has led to the use of flue-gas desulphurization in boilers, and improved burners to reduce the emissions of nitrogen oxides. However, this has increased costs, and reduced efficiency - thereby increasing CO_2 emissions.

Emerging technologies for generating heat and electricity from coal offer greater fuel flexibility, higher efficiencies and lower environmental impact than conventional technologies. A wide variety of specific technologies, based on fluidized-bed combustion and gasification, are still developing rapidly, often cross-fertilizing from earlier designs and lessons. Existing and proven forms of bubbling and circulating fluidized-bed plants have been widely installed to take advantage of their fuel flexibility, smaller scale and reduced local environmental impacts, especially for urban sites, often for applications which use both heat and electricity - cogeneration.

Second-generation technologies, especially those using pressurized fluidized beds or integrated gasification to power combined cycle and other advanced gas turbine systems, offer greater efficiency improvements and environmental benefits at potentially lower costs. These technologies are not yet commercially developed, but conceptual designs and some demonstration plants show a rapid trend towards reduced cost and better performance.

The progress of such 'clean coal' technologies will depend upon: the nature and timing of emissions constraints and penalties; the extent to which development and demonstration of more advanced systems receives financial support from governments and/or from coal industries seeking to diversify 'downstream'; aspects of utility regulation, for example with respect to cogeneration; and international support for technology transfer to encourage developing countries to use these cleaner and more efficient systems. Clean coal technologies appear destined to play a steadily increasing role in global energy supply, but the pace at which they penetrate the market will depend strongly upon such policy developments.

The world now uses more than 3,500 million tonnes of coal a year. Almost all of it is burned in plant using traditional technologies that date back at least sixty years - stoker-grate firing for small units, and pulverized-fuel firing for large units. The environmental impact of burning coal this way is now causing serious concern. It emits noxious combustion gases like sulphur and nitrogen oxides (SO_x and NO_x), precursors of acid precipitation; and international agreements and national standards in many countries now impose strict controls on these emissions. Moreover, carbon dioxide from fossil fuels is now recognized as a key factor in the 'greenhouse effect' of global warming and climate change; and coal releases more carbon dioxide per unit of energy than either oil or natural gas. Global negotiations are now seeking agreement on stringent control of carbon dioxide emissions; any effective agreement will have a significant impact on future use of coal.

In many countries a major use of coal is to raise steam to generate electricity, in traditional coal-fired power stations. But the efficiency of the steam cycle for pure electricity generation is low, usually well below 40%. Flue-gas desulphurization to control sulphur dioxide and selective catalytic reduction to control nitrogen oxides both reduce overall efficiency further; and this aggravates the emissions of carbon dioxide. Within the past decade, however, a widening range of new coal-use technologies has emerged. They can minimize emissions of sulphur and nitrogen oxides, while simultaneously achieving higher efficiencies to reduce emissions of carbon dioxide. Among the most promising concepts are 'fluidized-bed combustion' (FBC) and 'gasification', in a rapidly expanding variety of designs; some of the latest designs combine the advantages of both concepts.[1]

9.1 Coal-use technologies

In FBC, the combustion chamber contains a bed of inert particles like sand or ash, which is 'fluidized' by combustion air blown in from below, and heated to incandescence. Coal or other fuel injected into this turbulent glowing bed ignites and burns. Because the bed acts as a heat reservoir, FBC can burn almost any combustible material, including very low-grade fuel. Coal, of course, is not one substance but a wide range of substances, with widely varying characteristics; but an FBC unit with appropriate fuel-handling equipment can burn the entire range, from lignite to anthracite. It can also burn pit spoil, heavy oil, peat, wood, wood waste, refuse, sewage sludge - the list grows steadily longer. With fuel prices and supplies ever harder to predict, this fuel flexibility is one of FBC's most attractive attributes. The operating temperature of FBC, about 850°C, is low enough to minimize NO_x formation; staged injection of combustion air can reduce it yet further. If a 'sorbent' like limestone or dolomite is injected, most of the sulphur in the fuel reacts with the calcium in the sorbent and is retained in the bed as solid calcium sulphate.

FBC designs can be classified into three modes: *bubbling* (BFBC), *circulating* (CFBC), and *pressurized* (PFBC). BFBC is simple and compact, suitable for units of output up to some tens of megawatts of heat. CFBC is more complex, requiring a tall combustion chamber and heavy-duty cyclone; but it offers greater fuel flexibility and better control of SO_x and NO_x, and is suitable for units with outputs from tens to hundreds of megawatts. PFBC is the most complex of all, requiring a pressure shell and equipment for feeding fuel and sorbent and removing ash under pressure; but it offers good control of SO_x and NO_x, and its pressurized combustion gas can be fed directly into a gas turbine. Instead of requiring a premium fuel like natural gas, a combined cycle (CC) plant

---

1. Except where referenced below, for a fuller description of and commentary on the concepts outlined here see Walter C.Patterson, *Coal-use Technology in a Changing Environment*, Financial Times Business Information, 1990. For the latest information see, for instance *Modern Power Systems*, and *Coal and Synfuels Technology*.

based on PFBC can burn cheap low-grade coal, with an overall generating efficiency above 40%.

FBC of any kind requires sorbent for sulphur-trapping; a large unit, for instance, may require tens of thousands of tonnes of limestone per year. Extracting and transporting the sorbent may pose an environmental problem; so may disposing of the even larger quantity of sulphated sorbent. Tests indicate that the material is generally benign, and indeed that it may be usable as an analogue for light-weight concrete, or even gravel; but the specifications must be established for each particular fuel and sorbent at each particular site.

Coal gasification is another route to coal-fired CC, as 'integrated gasification combined cycles' or IGCC. Modern gasifiers can be classified into three types: *stationary bed, fluidized bed* and *entrained flow*. Each produces a fuel gas of moderate calorific value, lower than that of pipeline-quality natural gas, but entirely satisfactory as a fuel provided it does not have to be transported any considerable distance. Its sulphur content, mostly hydrogen sulphide, can be removed by conventional chemical clean-up, emerging as either elemental sulphur or sulphuric acid. Both are commercially marketable by-products. The fuel gas is then burned directly in a gas turbine, as the first stage of a CC plant, with an overall generating efficiency of more than 40%.

Different gasifiers offer different advantages and disadvantages. Some feed the fuel dry, others as a water slurry, affecting fuel flexibility and efficiency. Some require a large and costly 'radiant boiler' to recover heat from the gasification reaction. Those that do not recover this heat for use have lower efficiency. Some discharge ash as coarse agglomerate, some as granular slag. Both appear to be benign, and the slag can be used directly as aggregate, for instance for roadways. Current gasifier designs are mostly based on gasifiers originally intended to convert coal completely into 'synthesis gas' for manufacturing chemicals. Complete conversion of coal usually requires an oxygen plant, entailing extra capital cost and a significant fraction of the electrical output of the CC plant, effectively lowering the overall efficiency. Some recent gasifiers, specifically intended for power generation, use ordinary air, eliminating this efficiency penalty; the unconverted coal is burned to raise steam for the steam cycle in the CC plant.

All these technologies are evolving with remarkable speed, often cross-breeding between concepts to form hybrids that promise yet higher

Table 9.1 Comparitive performance of coal-use technologies for electricity generation: indicative estimates

Technology	Thermal efficiency %	Unit capital cost (1990 £/kW)	Generating cost (1990 p/kWh)
PF+FGD[*]	37	950	3.7
CFBC	38	885	3.5
PFBC	40	760	3.1
IGCC	42	780	3.1
Topping cycle	45	715	2.8
Natural gas CC	47	370	2.0-2.4

Note: [*]conventional pulverized fuel and flue gas desulphurization.

Assumption: 10% discount rate; coal at £1.7/GJ; natural gas at 24p/therm.

Source: J.Harrison, Coal Research Establishment, in presentation to British Institute of Energy Economists, London, meeting spring 1991, Chatham House.

efficiency, lower environmental impact and lower unit costs. CFBC, for instance, does not of itself much increase generating efficiency, although it avoids the loss of efficiency involved in flue-gas desulphurization (FGD). But a gasifier and gas turbine 'topping cycle' can be added, converting the plant to CC, burning the solid 'char' residue from the gasifier in the CFBC unit. Again, PFBC efficiency is limited because the fluidized bed operates at only 850°C, whereas a modern gas turbine can accept an inlet temperature of 1,260°C. Accordingly, fuel gas - perhaps from a gasifier - can be blended with PFBC combustion gas and burned in the gas turbine inlet, raising the temperature and the overall efficiency (see Table 9.1 for generic performance data). These technological permutations and combinations suggest that the potential for FBC and gasification can be extended significantly further.

Because the applications for both FBC and gasification vary widely, as will be described below, costs also vary widely from unit to unit. CFBC is already a commercial technology, in a hotly competitive international context; and manufacturers are reluctant to provide much financial detail. Recent turnkey plant, including steam-raising equipment, turbo-alternator and ancillaries, have been ordered at contract prices ranging from less than $1,400/kW to more than $2,000/kW. The wide disparity of costs arises from the particular plant configuration chosen, including the number of individual boilers and turbines. If the plant is to burn cheap waste fuel, the fuel-handling gear may be correspondingly expensive, to cope with awkward materials. Other features may likewise be particular to the given design.

Neither PFBC nor IGCC is yet fully commercial; cost data are thus at best indicative. The Tidd PFBC project in Ohio, replacing an existing steam system but retaining the existing steam boiler, had an estimated cost of $185 million, for an output of 80 megawatts of electricity, or about $2,300/kWe. The follow-up project, now in the design stage, is a 330 megawatt PFBC at the nearby Sporn plant, at a cost estimated to be $579 million, or $1,750/kWe. These are of course demonstration plants, conservatively designed and heavily instrumented; commercial units would be less expensive. On the other hand these prices are for repowering projects, retaining some of the original plant; a complete turnkey PFBC plant would cost correspondingly more. Shell has estimated that the unit capital cost of a 250 megawatt European IGCC plant like the one now under construction at Buggenum in the Netherlands would be some $1,910/kWe; they add that scaling up to 400 megawatts would reduce this to some $1,400-1,500/kWe.

The capital cost of these coal-use technologies is, and always will be, higher than that of plant of equivalent capacity burning oil or natural gas, simply because coal is a solid fuel and produces a solid waste. What matters, however, is the cost of the electricity and/or heat delivered. At the present low price of natural gas, coal-fired generating plant is hard-pressed to compete. Nevertheless, coal reserves remain much greater than natural gas reserves worldwide. The current enthusiasm for natural gas CC seems likely to accelerate depletion of natural gas reserves and exert strong upward pressure on prices, especially if supplies from politically volatile areas like the former Soviet Union and the Middle East are jeopardized. Coal prices are unlikely to rise so much;

coal reserves are too large, suppliers too numerous and reserves too widely distributed. As natural gas prices rise, advanced coal technologies will look increasingly attractive.

9.2 Primary market areas

Advanced coal-use technologies may be retrofitted to existing boilers and furnaces, or may serve as the basis for new greenfield installations. Some dozens of FBC retrofits are now in operation, mostly BFBC but also including a few large CFBC units and three demonstration PFBC units. An FBC retrofit can replace the bottom of the combustion chamber alone, or the whole of the combustion chamber, or the whole steam-raising plant including heat-exchange surfaces. Experience suggests that attention to site-specific details is essential: otherwise the nominal capital savings from retaining part of the original steam-raising plant may be offset by operating and maintenance problems arising from the difficulty of matching the original plant - fuel-handling gear in particular - to the retrofit.

Until recently, the size range of FBC units available made them suitable primarily for industrial boilers and furnaces and for combined heat and power. In the late 1970s and early 1980s, hundreds of small BFBC boilers and furnaces came into service, both new plants and retrofits of existing plants. Since the early 1980s, however, the focus has shifted to CFBC, in ever-increasing sizes. CFBC allows greater fuel flexibility and better control of SO_x and NO_x. The low emissions allow CFBC plants to be sited in urban areas without major detriment to the environment. Accordingly, many of the more than 100 CFBC units now in service worldwide, in sizes from tens to more than 400 megawatts thermal, are cogeneration plants. They raise steam to generate electricity with back-pressure or pass-out turbines that also supply process steam for local industry, or steam or hot water for local district heating. Such applications also achieve overall fuel efficiencies better than 80%. Some of the latest CFBC units, however, are for pure electricity generation, in capacities of 150 megawatts of electricity and above, with reheat steam cycles; Electricité de France is involved in design studies for a CFBC unit with a capacity of 250 megawatts of electricity. Such plant can use cheap low-grade fuel of a wide range of specifications with minimal preparation, while meeting prevailing standards for SO_x and NO_x. Nevertheless, using a pure steam cycle means that their efficiencies, albeit better than those of traditional coal-fired plant with flue-gas desulphurization (FGD), are still likely to be less than 40%.

Higher efficiencies for pure electricity generation using coal can be achieved only by combined-cycle operation - that is, by PFBC or IGCC - or by technologies still at the laboratory stage, like coal-fired fuel cells and magnetohydrodynamics (MHD). The 330 megawatt Sporn PFBC plant may get the go-ahead in the US by 1993. The first utility-scale IGCC plant, the 250-megawatt Buggenum unit, is now under construction in the Netherlands; and at least two others of comparable size have been ordered in the US, in Massachusetts and Florida. Progress will be studied closely. A number of utilities are carrying out feasibility studies of utility-scale PFBC and IGCC units, with a view to ordering plant later in the 1990s. Companies in Britain, Germany, France and the US are also pursuing the possibility of adding a coal gasifier, gas cleanup and gas turbine - a 'topping cycle' - to existing coal-fired steam plant, whether traditional, CFBC or PFBC. This will increase its efficiency and reduce both SO_x and CO_2 emissions per unit of electricity sent out.

Until recent years utilities sought higher efficiency simply by increasing the size of steam plant, to 1,000 megawatts or larger. This trend has run its course; and utilities are now more interested in plant of moderate size, up to perhaps 400 megawatts. Such plant is easier to site; much of it can be shop-built, reducing site problems; it is likely to be more reliable in operation, and require less backup capacity; and it can be built and brought on stream in three years or less, alleviating problems of forecasting for system planning. Advanced coal-use technologies fit well into this new utility philosophy.

A recent census by IEA Coal Research catalogued more than 1,300 coal-fired electricity plants around the world.[2] Many of these will be candidates for retrofitting or replacement within the coming fifteen years; advanced coal technologies will undoubtedly figure prominently in the plans. Several developing countries, among them China, India and Brazil, have major coal reserves and are striving to expand their electricity systems. Commentators stress the need for cooperation, to enable such countries to install the cleanest and most efficient coal-firing technology available; but problems of technology transfer remain challenging. For most developing countries shortage of capital for

---

2. A. Mannini et al, *World Coal-fired Power Stations*, IEA Coal Research, HMSO, September 1990.

investment is a daunting problem. The major engineering firms now active in advanced combustion technology cannot hope to sell plant on straightforward commercial terms; developing countries simply cannot afford them. Opportunities for joint ventures, licensing on 'soft' terms, or even 'build-operate-transfer' deals will need to be explored. International aid and development agencies like the World Bank will have a crucial role to play. Forthcoming global environmental agreements like a carbon dioxide convention may have to incorporate suitable measures to offer technical assistance to developing countries.

The fuel flexibility of many of the advanced combustion technologies offers a way to overcome the often low or unpredictable quality of fuels available in developing countries. In the longer term, and perhaps especially in developing countries, this same fuel flexibility will allow a gradual transition away from fossil fuels like coal, to the use of non-fossil fuels like biomass - wood and wood waste, bagasse from sugar cane, and similar cellulose materials - whose combustion does not add fossil carbon to the atmosphere. Biomass gasification combined cycles, for instance, could become an increasingly important generating technology, utilizing plant originally installed to burn fossil fuel.[3] Biomass can be gasified more easily than coal, often in the same gasifier; indeed some modern gasifiers were designed initially for biomass. The concept is already technically feasible; but its economic status and environmental impact will depend on organizing a sustainable and environmentally acceptable supply of suitable biomass fuel. In the meantime, coal will continue to be an essential component of the evolving fuel mix.

9.3 History

Until the mid-1970s almost nobody was interested in technological alternatives to stoker-grate and pulverized-fuel firing for coal. From the 1950s onwards coal itself had faced increasing competition from cheap oil and natural gas. Energy research and development, especially that funded by governments, was devoted almost exclusively to nuclear power. Early efforts to develop coal-firing for gas turbines failed. The engineering development that did affect coal-fired power plant was

3. Eric Larsen et al, 'Biomass Gasification for Gas Turbine Power Generation', in T.Johansson et al (eds), *Electricity: Efficient End-use and New Generation Technologies and their Planning Implications*, Lund University Press, Lund, Sweden, 1989.

mainly on the steam cycle itself, scaling up unit size and increasing steam conditions in pursuit of higher steam-cycle efficiency.

The concept of FBC emerged in Britain and the US at the beginning of the 1960s, but attracted essentially no interest from electricity suppliers. The ability of FBC to trap fuel-borne sulphur during the combustion process itself had been identified by 1967. In 1968 Britain's National Coal Board proposed to build a coal-fired FBC demonstration power station at Grimethorpe, in Yorkshire; but the then Central Electricity Generating Board instead opted for a nuclear station now known as Hartlepool. In 1968 British engineers proposed the concept of PFBC, and the following year built a test rig that remained the world's largest through the following decade, carrying out much of the fundamental research that proved the feasibility of the concept. They worked, however, mainly under contract to clients in Sweden and the US, as successive British governments ignored the technology in pursuit of ever more trouble-prone nuclear power.

The first FBC power plant was a BFBC retrofit to a small old coal-fired station at Rivesville, West Virginia in the early 1970s. It was done on a shoestring, and the cost-cutting led to endless problems that culminated in shutdown before the end of the 1970s. Other early units experienced erosion and corrosion of boiler tubes, clogging of air distributors and frequent difficulties with coal feed. From the mid-1970s onwards, however, a series of industrial demonstration boilers and furnaces in the US and Britain confirmed technical feasibility; some are still operating today.

Meanwhile, however, CFBC emerged, initially from private industry in Scandinavia and with government support in Germany. The first demonstration plants were operating by 1980, and the technology thereafter took off with remarkable speed with scarcely a technical hitch, evolving virtually from plant to plant, increasing in size and versatility while meeting stringent emissions standards. Within a decade CFBC industrial boilers and cogeneration plants, district heating plants, and even utility-scale power plants have come into service in more than a dozen countries, from a lengthening roster of major engineering manufacturers. Questions of longer-term maintenance and reliability of course remain to be answered; occasional tube leaks and problems with refractory are being closely watched. But CFBC is already clearly a commercial technology, with a number of manufacturers competing hotly for new contracts across the world.

A tripartite agreement between Britain, Germany and the US in 1975 led to construction of the first large PFBC rig, at Grimethorpe in Britain. Many technical problems were encountered, especially relating to erosion of in-bed tubing; but many of them were solved, contributing substantially to advanced testing of PFBC configurations. Work in Sweden confirmed the feasibility of the concept; and three demonstration plants - at Tidd in the US, Escatron in Spain and a two-unit plant in Stockholm - are now in the commissioning stage and appear to be performing well. At least three engineering manufacturers now have significant PFBC programmes.

Gasification has a much longer history than FBC. But modern high-throughput, high-conversion gasifiers have been around for less than two decades. Most of the work in the early 1970s was carried out by oil and chemical companies seeking a way to produce coal-based chemical feedstocks or pipeline-quality 'synthetic natural gas'. Active interest in IGCC for electricity generation dates back only to the beginning of the 1980s, after the first prototype and demonstration gasifiers for feedstocks were already coming into service. The Cool Water project, a 100 megawatt IGCC plant in California, was a remarkable engineering success, starting up in 1984 ahead of schedule and under budget, and performing even better than expected - both more flexibly and more reliably. The Deer Park and Plaquemine demonstration plants both started up in 1987, and have likewise performed very well, spurring interest not only in the particular gasifiers involved, but also in IGCC itself. Quasi-commercial IGCC plants are now under construction in the Netherlands and the US; further units are planned, and more gasifier designs are under active development. A related concept, the gasifier topping cycle, is also reaching prototype stage in Germany and the US. Britain's topping cycle project was granted a modicum of government funding in 1991; but coal technology research remains the poor relation in Britain.

9.4 Pressures for increased uptake

The future use of coal itself faces many uncertainties. In recent history, coal has been used mainly to supply industrial process heat and steam, electricity, and district heating. In the industrial world all of these markets are under threat from both ends. Higher end-use efficiency of buildings and process plant will reduce the need for fuel of any kind to deliver the final comfort, light, process output and other 'energy services'. Governments around the world enthusiastically endorse the concept of increased efficiency; but many are laggard in implementing the practical measures already available, as described elsewhere in this book. Because coal is a comparatively awkward fuel to use, it must compete on price and availability against oil and gas in its traditional markets in the industrial world, and also in the expanding fuel markets of the developing countries. The advanced coal-use technologies, with their flexibility, higher efficiency and lower environmental impact, will help to reduce the gap between coal and competing fuels that have hitherto been more convenient.

The size of this gap of course depends on the actual and anticipated price of oil and natural gas where they can be used as alternative fuels to coal. In the present climate of uncertainty about future oil and gas prices, with dramatic changes not only year by year but week by week, no confident prediction is possible. Plant operators can seek long-term fuel supply contracts whose price structure includes appropriate escalation clauses; but the present volatility of the market suggests that long-term contracts may routinely link one winner and one loser. Against this background, the relative stability of coal prices is itself an attraction, especially if it can be coupled with coal-use technology whose fuel-specifications are undemanding and flexible. Moreover, in many countries coal is a domestic resource, whose security of supply may enhance its attractions against imports - provided the coal can be burned cleanly.

In general, current standards for emissions of SO_x and NO_x and for efficiency, where they are imposed, will enhance the attraction of advanced coal-use technologies compared to traditional technologies. To be sure, a move to these technologies will entail major investment in new plant, although as the technologies mature their unit capital costs are already declining, and this decline will probably continue.

But some standards - for instance those for NO_x in Japan - are already so stringent that even CFBC and IGCC plant may need to incorporate additional NO_x control such as selective catalytic reduction (SCR). Depending on the particular fuel used and the standard in effect, SO_2 control may be easier for IGCC than for CFBC and PFBC. CFBC and

PFBC require sorbent extraction and generate solid waste of limited economic value at best, whereas IGCC does not require sorbent extraction and generates marketable sulphur or sulphuric acid.

Carbon dioxide emissions limits will pose a much more serious problem for all fossil-fuel technologies, coal most of all. So long as fossil fuels are used at all, higher overall efficiency will become crucial. This implies, for instance, a steady expansion of cogeneration - for which the advanced technologies are especially well suited. Their fuel flexibility will also open the way to a progressive shift from fossil to biomass fuels with no net carbon emissions, in the same or similar plant.

For coal suppliers, this presents an important corollary. Higher efficiency means using a lower quantity of coal to deliver the same service. Coal suppliers may find their markets smaller than expected, and intense competition may reduce profit margins on coal sales even further - especially once biomass fuels become significant. One way for suppliers to counter this threat to their business will be for suppliers themselves to diversify downstream. Some coal suppliers are already actively involved in developing technologies to use their coal. They could expand these activities, for example, by building and even operating the power stations that use their coal. Profits from selling electricity could be substantially higher than those from selling coal; and the higher efficiency and lower environmental impact of the new technologies could make them the key to successful downstream diversification. If coal suppliers themselves seize this opportunity, the role of the new technologies could expand rapidly.

9.5 Constraints on uptake

The worldwide recession in industrial countries has created an unpropitious climate for investment by manufacturing industries and utilities. Demand for goods and services, and for the electricity and heat needed to provide them, is low, and interest rates are high. Orders for new plant of any kind, including advanced combustion plant, are unlikely to pick up until the world economy recovers significantly. Developing countries likewise face continuing acute difficulty with access to capital for industrialization. They also lack skilled people to build and operate new plant.

Even if new plant is contemplated, buyers are likely to look sceptically at advanced coal-use technologies if they are confident about future supplies and prices of oil or gas. Their scepticism may be reinforced by doubts about the long-term reliability and performance of the new technologies, and by questions about further tightening of environmental constraints on the use of coal.

9.6 Policy mechanisms and options

As indicated above, in sum, policy measures that will affect the rate and scale of moves to new coal-use technologies include:

* emission controls and standards;

* carbon dioxide protocols and emissions targets;

* the extent and focus of government financial support for demonstration plants;

* downstream diversification by coal suppliers;

* the impact of local planning controls;

* fiscal incentives;

* electrical utility regulation, especially concerning the role of cogeneration, district heating and independent power production;

* steps towards effective technology transfer arrangements between industrial and developing countries.

9.7 Likely and potential impact

The impact of emerging coal-use technologies in the 1990s will be determined by the implementation or otherwise of policy measures like those outlined in the preceding section. The likelihood of this differs markedly from country to country. Some national governments, notably the US and Japan, already have active and substantial programmes of support for new coal technologies. Curiously enough, although the US of course is a major coal producer, Japan's domestic coal industry has almost disappeared. Other countries - of which Britain is a prime example - have domestic coal industries but give them little if any support, in either policy or finances. Still others, like Denmark, are actively hostile to coal in general, for environmental reasons, regardless of the technology employed. Developing countries with major coal industries, notably China, India and Brazil, are aiming for rapid expansion of their energy systems, and have some interest in the new coal technologies; China in particular has some 2,000 small and basic FBC units, built between the mid-1960s and the early 1980s. But these countries continue to focus on traditional coal-use technologies, abetted by engineering firms in the industrial world and by international funding agencies like the World Bank.

The prospects for the new coal technologies are thus widely disparate in different places. But the rate of development over the past decade, against a generally unpromising economic and environmental background, has been remarkable. Historically, coal has played a central role in the evolution of industrial society. It continues to be abundant and widely available, from many suppliers, including domestic suppliers in many countries that use coal. Unlike oil, coal is not subject to the vagaries of an international cartel; unlike natural gas, coal is not dependent on supplies from areas that may be politically volatile. In these important respects coal is therefore a more reliable fuel than either oil or gas. From now on, however, the world seems bound to insist that coal clean up its act. In the next decade FBC and gasification, especially for cogeneration and combined cycles, will offer a burgeoning catalogue of opportunities to do so. They will also help to build essential bridges towards a future of high efficiency and low emissions, whatever the fuel.

Wind Energy

Wind energy for power generation has developed very rapidly since the mid-1970s to being a commercially-available technology with substantial operating experience. The economics vary greatly according to the siting and financial conditions, but it is already competitive without subsidy in the most favourable conditions and the technology is continuing to develop.

The major developments have occurred in Denmark and California in response to government policies which supported renewable sources and encouraged private sector power generation. Lessening support led to contraction and cost-cutting in 1986-8 but interest broadened again in the late 1980s, and now includes other European countries and other states in the US, and some developing countries. Industry projections suggest that at least 10,000MW will be installed worldwide by the year 2000.

The rate of growth will depend strongly on government policies. One critical factor is the value of wind-generated electricity, which depends on: utility financing conditions; the degree of subsidies to competing sources and/or internalization of their external impacts; and the regulations governing independent power production. Direct support for wind energy to overcome initial hurdles and reflect environmental benefits over other options, protection of independent wind generation against utility monopoly powers, streamlining of planning procedures, and national wind resource surveys are also important. Capital support for transferring the technologies to developing countries could enable rapid growth in these regions. Experience demonstrates the importance of designing policies carefully: some support has unwittingly encouraged bad engineering and siting.

Potential wind energy resources are very large, and in some major countries exceed total electricity demand. The most important constraints are those of visual impact, limited feasible high-windspeed areas, and in some cases mis-matches between system demand and the availability of wind energy. It is too early to know how such factors will limit wind energy and many other factors remain uncertain, but wind energy has clear potential to become an important component in power generation during the first few decades of the next century.

The use of wind energy for mechanical work dates back thousands of years. Traditional windmills once numbered ten thousand or more in Holland and as many again in England, but they were very inefficient and were inevitably displaced by the rise of steam power in the 18th and 19th Centuries. Despite the decline in numbers of traditional windmills in industrialized countries, small wind turbines of simple design are still used for water pumping in areas remote from electricity networks, especially in Australia, Argentina and the US, and for electricity generation in conjunction with battery storage. Wind energy for island supplies, to avoid high-cost fuel imports, forms another important use.

However, few energy analysts have taken wind energy for bulk power generation seriously. But wind technology has developed very rapidly during the 1980s, and wind power is already competitive for bulk power generation in some conditions and systems. In addition to traditional uses for isolated applications, by 1990 there were more than 20,000 electricity-producing wind turbines connected to mainland electricity grid systems, mostly in Europe and North America. Wind energy in 1990 supplied about 2% of electricity in California and Denmark - the Danish figure is projected to rise to 10% by 2000 in the official Danish environmental energy scenario.[1]

---

1. M.Grubb and N.Meyer, 'Wind Energy: Resources, Systems and Regional Strategies', in Thomas B.Johansson, Henry Kelly, Anulya K.Reddy, Robert H.Williams (eds), *Fuels and Electricity from Renewable Energy Sources*, Island Press, Washington DC, forthcoming 1992.

The technology for wind energy is still developing rapidly, and during the 1990s wind energy could enter a phase of major expansion. The total physical wind energy resource is many times global electricity demand. Even taking into environmental, land-use and system constraints, wind power could eventually supply more than 20% of global electricity demand.[2] This chapter examines the current status of wind energy technology, the potential for expansion and the factors which will determine this, the policies which have brought it to its current stage and those which will affect its future progress.

10.1 Characteristics of wind energy

Modern wind turbines differ from the old sail mills in ways. Most of those for generation on large power systems have two or three aerofoils, mounted on a horizontal rotor which is attached to a gearbox and generator at the top of a single tubular tower. Commercially-available machines range in size up to 35 metres in height (hub height above ground level) with capacities up to 500kW (Figure 10.1). There have been experimental machines several times this size. There are many other configurations: some have even been constructed with just one blade with a counterweight, whilst others use vertical blades mounted around a vertical axis.

The primary attractions of wind energy are its environmental advantages and the small unit size. The main obstacles are the relative youth of the technology in its modern form, its capital intensive nature, the visual impact of wind turbines and consequent difficulties in siting large numbers, and (especially for remote applications) the intermittency of the winds.

The environmental attractions of wind energy are as for most renewable sources. It produces no solid or liquid wastes, and no gaseous emissions. It requires no external fuels and so avoids the environmental problems of fuel extraction and transport, as well as the economic dangers of fuel supply interruption and price hikes. Siting is not dependent on the availability of cooling water, and land can be shared with other applications such as farming.

Potential environmental drawbacks include electromagnetic interference, noise, and visual impact. To avoid interference with TV and other transmissions, wind turbines have to be sited away from main

2. ibid.

Figure 10.1 Modern commercial wind turbine

Source: Wind Energy Group, Greenford, UK.

transmitters, and not on line-of-sight of microwave communications (eg. radar and field telephones). Reception interference can also occur, especially in areas of already low quality, but a local relay station to amplify the signal, or cable connections to affected houses, can be installed at a small fraction of the windfarm cost. Electromagnetic interference is thus not considered a significant obstacle, and no sites in Denmark have been refused on these grounds.[3]

The noise from wind turbines places a limit on how close to houses they can be sited. There is considerable variation according to machine design and location.[4] Problems with unusually noisy machines have led to their withdrawal from the market, but noise still constrains the available sites; regulations in Denmark restrict noise levels at dwellings to 45dB, a level which is 'greater than in a bedroom at night, but less than in a house during the day'.[5] Current technology requires a distance of about 300m for single machines, or up to 500m for a windfarm to meet these specifications.[6] Significant numbers of bird deaths have not been noted in Europe, but bird strikes of some species has emerged as an issue in the US.[7]

The most common objection raised against wind energy is that of visual impact. For wind energy to make a substantial contribution, thousands of turbines would be required, and so they would have to be visually acceptable. Visual impact as it constrains the available wind resource is discussed further below.

The relatively small unit size of wind turbines compared with most power plants opens the technology to relatively small investors and enables very rapid installation, usually within a few months of signing contracts. It allows production line methods to be used, allowing

---

3. ibid.
4. Difficulties in measuring and interpreting noise data are discussed in A.Robson, 'Environmental aspects of large scale wind power systems in the UK', *IEE Proceedings A*, Vol.127 No.5, IEE, London, 1980; issues of *Windpower Monthly* report occasional public complaints against particular installations.
5. European Wind Energy Association, 'Wind Energy in Europe - Time for Action', EWEA, Oxford, October 1991.
6. Grubb and Meyer, op.cit.
7. 'Birds ruling affects 8000 turbines', *Windpower Monthly*, January 1991; 'Bird deaths study reveals true cause for alarm', *Windpower Monthly*, May 1991. Recent studies suggest about one large bird-of-prey may be killed each month by the Altamont Pass windfarms (Elliot, private communication).

fixed-price contracts for standardized products, with the primary risk being borne by the manufacturer rather than the purchaser.

Furthermore, when problems do occur, it is usually possible to replace faulty components relatively quickly. Modern wind turbines therefore tend to be technically available more than conventional plant; the best wind farms have registered more than 95% annual availability. Such levels are, however, only achieved by having maintenance staff readily available to correct failures, which would add to maintenance costs particularly if there only a few small installations in a given area.

Winds fluctuate on all timescales. The value of wind on a large power system does not depend much on having reliable output at times of peak demand, because there is still a statistical contribution to reliability and because plants such as gas turbines can provide very cheap backup capacity.[8] Also, for moderate capacities the power fluctuations are simply drowned out in those of the electricity demand, so that operational penalties are negligible.[9] But the broader relationship between wind output and demand can be important, depending partly on the existing mix of plant. An inverse relationship can substantially reduce the wind value (eg. southern California) because much of the energy displaces cheap baseload plant. Conversely, correlation with system demand on a seasonal (eg. northern Europe) and/or daily (eg. northern California) basis boosts the value of wind energy because it then displaces more expensive fuels. In such systems, the wind energy is usually at least as valuable as that from conventional stations until the contribution exceeds 5-10% of system demand.[10]

If the contribution were to grow beyond such levels, the value of the wind energy would decline - further wind turbines would save little or no additional conventional capacity, their output would displace successively less expensive fuels, the operational penalties would increase, and at very high penetrations, occasions when wind energy

8. M.J.Grubb, *The value of variable sources on power system, IEE Proceedings C*, 138:2, Institution of Electrical Engineers, London, 1991, and references.
9. ibid; the former Central Electricity Generating Board of England and Wales estimated an operating penalty of 2% of fuel savings for their planned windfarms (*CEGB Statement of Case to the Public Enquiry on the planned Hinkley Point C nuclear reactor, 1988*).
10. M.J.Grubb, 'The economic value of wind energy at high power system penetrations: an analysis of models, sensitivities and assumptions', *Wind Engineering*, Vol.12:1, 1988, pp.1-26.

would have to be discarded to avoid exceeding the maximum the system could safely absorb would become economically significant. On most large integrated systems there are no technical reasons why wind energy cannot supply more than half the electricity demand even in the absence of storage. But even if the costs of the technology decline substantially, and sufficient resources are available, the declining economic value would probably limit the economically plausible contribution of wind energy to 20-40% on most systems in the absence of storage or hydro power.[11] However, this 'integration limit' is sufficiently high that in practice system integration will not be a significant issue for the foreseeable future in most countries.

Depending very much on local network conditions, the lower 'quality' of power from some wind turbines can create operational complications, or require additional reactive power on the system to avoid local voltage fluctuations, which may add a few per cent to the wind turbine costs. For remote applications, intermittency and lower power quality can be a more serious problem. A single machine, or group clustered on one site, provide much more variable power than if many turbines are spread across a grid, and small power systems are usually less able to absorb variable sources. In isolated applications (excepting some pumping) storage may be required - more than for solar sources, because the variation is less regular. Also, the diesel plant often used for generation on small power systems can be inflexible in operation: the fuel costs, and hence potential savings, are high, but when wind energy is included, sophisticated control strategies may be required to avoid large operational losses.[12]

10.2 Wind energy developments

Most of the underlying technologies on which wind turbines rest are many decades old, and there have been several periods of research interest in 'modern' wind energy since the beginning of the century. Despite this, the main developments did not start until the oil price shock of 1973.

The subsequent evolution of wind energy for grid supplies has occurred in four distinct phases. A range of government programmes, notably in

11. ibid.
12. N.H.Lipman et al, *An Overview of Wind/Diesel R&D Activities*, Rutherford Appleton Laboratory, Oxford, UK, 1989.

the US, Sweden, Germany, and Canada began between 1976 and 1981, aimed at developing very large turbines and understanding the underlying technology. While much was learned, most projects ran into substantial technical problems and high costs.

The second phase, from 1982-85, was dominated by the development of a market for small and medium-sized machines in the US. A favourable regulatory regime combined with generous Federal and State tax incentives in the early 1980s made wind energy in some areas - particularly California - an attractive private investment even at the then high costs. Installation rates in California rose from 10MW in 1981 to 60MW in 1982, and trebled in each of 1983 and 1984 to a level of 400MW a year. The cumulative investment in Californian wind energy by 1986 totalled about US$2,000m, with the value of the electricity generated put at US$100m a year.[13] The total cost of the programme net of fuel savings (discounted back to 1980) has been put at about US$500m.[14]

These developments had major drawbacks as well as benefits. Many of the early machines installed were of poor quality, and some broke in the first season of operation - as one committed manufacturer complained, many of the early companies knew more about tax management than engineering. It was indeed claimed that at one time, high taxpayers could profit by investing in wind machines even if they never worked. Machines were often sited carelessly and very densely, and some were sold on fraudulent promises. 'These developments will destroy wind energy' was the reaction of the Electric Power Research Institute's wind project manager in 1983, and lines of ugly, motionless machines certainly damaged the reputation of wind energy.[15]

But with the market base and finance of California, several companies invested heavily in wind energy technology and gained rapid experience. The results over this period were: a doubling in the mean size of commercial units to over 100kW; major improvements in machine performance; and a rapid fall in capital costs, from US$3,100/kW in 1981

---

13. P. Gipe, 'Maturation of the US wind industry', *Public Utilities Fortnightly*, 20 Feb. 1986.
14. Alan Cox, Carl Blumstein, Richard Gilbert, 'Wind power in California: a case study of targeted subsidies', Universitywide Energy Research Group report UER-191, University of California, 1989.
15. F. Goodman, private communication, 1983. For discussion of the programme see Cox et al, ibid.

to an estimated US$1,250/kW average in 1986 (historic prices).[16] Figure 10.2 shows some measures of the development of wind energy in California during the 1980s.

The third phase of wind development, from late 1985, was dominated by the removal of US tax credits and the fall in oil prices. This greatly tightened the market at a time when several large companies had put substantial capital into new machines. The resulting pressure led to further cost and price cuts, this time driven by market pressure as much as technological change. In combination with the decline of the dollar (which greatly reduced revenues for non-US manufacturers in the US market) a number of companies went bankrupt and others merged.

The fourth phase, emerging in the late 1980s, has been one of resumed expansion, with an initial focus on Europe. The Danish target to install 100MW of utility wind generation by the end of the decade was met with a rush in 1988 and 1989, and with more than this being installed by non-utility generators. Tens of megawatts were installed in the Netherlands by 1990 in pursuit of a target of 100MW by 1992. In 1990 Germany declared a goal of 100MW by 1995, a target which was doubled a few months later; about 50MW was expected to be installed during 1991.[17] In 1988 the UK announced plans for its first three windfarms, totalling 24MW, proposals which were joined by other initiatives under the 'non-fossil quota' of the newly privatized electricity system in 1990. In 1991, the scale of activity expanded dramatically, with about 200MW of wind energy contracts being admitted under the non-fossil scheme.[18] Along with this, the turn of the decade saw renewed installations in the US (Figure 10.2), and activity in India and a number of other developing countries.

The technological trend during this fourth phase to date has not been one of falling machine prices - which at the beginning of the period were driven by market pressures to commercially unsustainable levels - but of steadily improving performance and increased size, yielding higher output per site and lower on-site costs. Most commercial machines in 1985 were under 100kW; by 1990 most manufacturers had units of 300-400kW. The poor performance of machines from the early 1980s was by the late 1980s giving way to availabilities of 95% and capacity

16. Gipe, op.cit.; *Wind Directions*, March 1987.
17. *Windpower Monthly*, various issues.

Figure 10.2 Measurements of wind energy developments in California, 1982-90

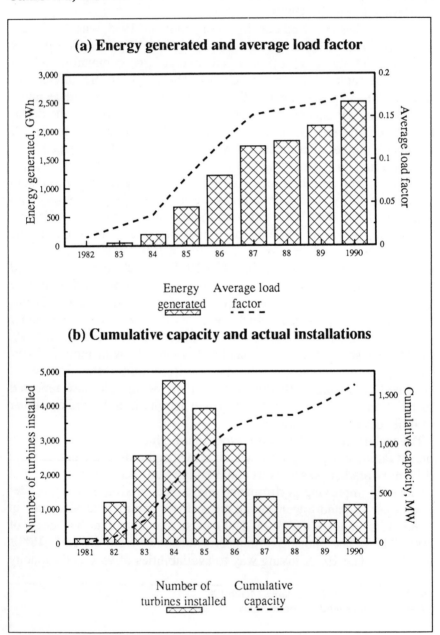

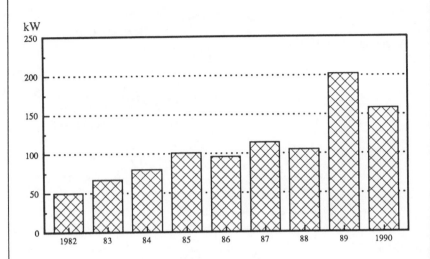

(c) Average unit size

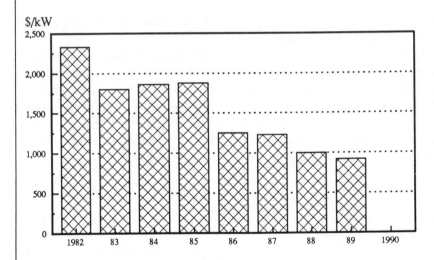

(d) Average capital cost

Source: P.Gipe and Associates.

Figure 10.3 Distribution of wind energy generation in 1990

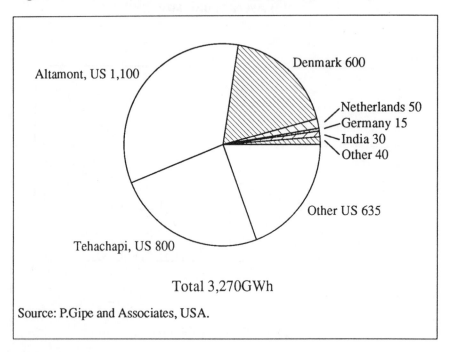

factors (average output as a fraction of peak power) up to 30%, double the levels five years earlier.

In 1991, Californian windfarms dominated total wind energy generation, as illustrated in Figure 10.3. The contribution of wind energy is still small compared with total power supplies; in California and Denmark, wind turbines in 1990 supplied about 2% of electricity, elsewhere the contribution was negligible.[19] Yet the demand for wind energy is now large enough to support a substantial industry and to prompt investments to drive the technology further.

18. The cited figure was 82.4MW of Declared Net Capacity, which roughly reflects average output taking into account the variability of the wind.

19. The Tehachapi and Altamont passes are both in California and provide the wind generation of 1,900GWh illustrated in Figure 10.3, as is most of the 'other US' generation shown in Figure 3, giving a total of about 2,500GWh, or 1.9% of the total Californian electricity generation of 132,000GWh in 1989. The Danish figure of 600GWh wind energy is 2.3% of total Danish electricity generation in 1989 (26,000GWh).

Improvements are expected through refinements of the relatively simple fixed-speed technology which currently dominates wind energy markets. But the biggest gains are likely to come through developments which overcome the inherent limitations of current technology. Having recovered from the recession of 1986-88, wind markets now appear in a position to move towards a new generation of wind turbines based on more advanced approaches, and many avenues are being explored. One development attracting great interest is the use of power electronics to produce a constant frequency output of high quality whilst allowing the rotor speed to vary. This gives many benefits, most notably an increase of perhaps 20% in energy capture as compared with fixed speed machines. A prototype developed between US utilities and the US Windpower company is stated to exploit these benefits with little extra capital cost.[20]

10.3 Economics of wind energy

The thread of continuity between the US 'windrush' and the developing European markets means that the critical technological lessons from the US are embodied in current machines. The dominant technology remains the relatively simple 3-bladed, fixed speed machine, but much improved blades, controls and other components have steadily boosted the performance as indicated above.

It is hard to generalize about the economics of wind energy. Machine costs vary according to the size and specification; costs per kW have proved a poor guide as they can readily be boosted by installing a bigger generator without yielding much extra energy, so it is now more common to quote costs per square metre of swept area.[21] The costs associated with installing a machine on a site vary with the location (roughness of the terrain, distance from the grid, etc) and can add anything in the range 15-50% to the capital costs. Maintenance costs and land costs also vary with the site. The return depends upon the energy output, which is very

20. EPRI Journal, June 1990; D.Cavallo, D.Smith and S.Hock, 'Wind Energy: Technological and Economic Aspects', in Johansson et al, op.cit.
21. 'Over-rating' of machines - installing over-sized generators to reduce apparent cost/kW - was a problem in early wind markets, and where government incentives depend on the capacity installed; this contributed to the low capacity factor of early machines. For machines optimally rated for a good site, the peak output is typically about 0.5kW per m^2 of swept area, in which case the costs in $/kW are roughly twice those in $/$m^2$.

Table 10.1 Complete costs of Danish windfarms installed in late 1980s

Machine size:	100-300kW
Average machine cost:	£200/m^2
Total installed cost:	£264-350/m^2
Average annual maintenance cost:	1.3% of capital cost
Average annual other cost	1% of capital cost

Note: converted from ECU at £1=ECU1.4

Source: P.Nielsen, 'Wind Energy Activities in Denmark', *Proc International Wind Energy Energy Conference*, Madrid, September 1990.

sensitive to the windspeed at the site (the energy available is proportional to the cube of the windspeed, so that a 10% difference in windspeed alters the energy available by 30%), and the value of the energy varies considerably between systems according to other generation costs and arrangements with the utilities. Finally, because the initial capital accounts for most of the costs, the overall economics are very sensitive to financing conditions: a high interest rate (or high discount rate, or high required rate of return) will greatly increase the cost of electricity. There is no simple answer to the question 'is wind energy economic?'

Nevertheless existing experience gives useful indications. Table 10.1 shows results of a study by the Dutch utility ELSAM of the total costs of seven windfarms built in the late 1980s, totalling 37MW. The implications of such figures for generating costs, adjusted for different siting regimes and different discount rates, are shown in Table 10.2. For typical Danish conditions with utility discount rates of 8%, the energy costs are in the region 2.9-5.4p/kWh depending primarily on windspeeds. At the best sites in the UK, using the 5% discount rate of the former

Table 10.2 Costs of wind-generated electricity from currently available technology

Type of site	Wind speed at hub height (m/s)	Installed cost £/m^2	Energy cost p/kWh at discount rate:			
			5%	8%	10%	15%
Fairly good	6.5	286	4.4	5.3	5.9	7.7
Good	7.5	357	3.6	4.3	4.8	6.3
Very good	8.5	428	2.4	2.9	3.3	4.3

Note: converted from ECU at £1=ECU1.4.

Source: European Wind Energy Association, 'Wind Energy in Europe - Time for Action', EWEA, Oxford, UK, October 1991.

public utility, costs would be under 2.5p/kWh, consistent with estimates made in 1988 by the utility.[22] In the radically altered conditions of privatized electricity in the UK, seeking 10-15% rates of return, energy costs are 30-70% higher.

With these costs, wind energy can already be competitive at exceptional sites, but in general these costs have not been sufficient to enable wind energy to compete unaided in the European electricity market even in countries like Denmark with relatively good resources and high electricity prices. Commercialization to date has depended upon some degree of tax credits or subsidy. Moves to reflect the external costs of existing electricity generation in the price paid for wind-generated electricity could change this picture (Chapter 2), particularly if combined with other policy developments as discussed below. But much will also depend upon further lowering wind energy costs.

As noted above, many technical avenues for lower capital costs and increasing output are being explored. Drawing on these developments

22. CEGB evidence to the Hinkley Point C reactor inquiry, op.cit., estimated the energy from their planned windfarms to cost 2.26p/kWh.

and the inevitable decline of costs with increased production, Table 10.3 shows past and projected dollar costs of wind energy under various conditions.[23]

These projections indicate continuing reductions in energy costs which in favourable conditions will make wind energy financially attractive compared with most conventional sources, even without accounting for external costs. In privatized electricity markets, the higher capital charges make the position less clear, though the ability to install windfarms in a matter of months rather than years is a clear benefit over conventional power stations because of the lower risks and interest accumulation before the first returns. In such markets, the dominant competition is likely to be not coal or nuclear but the gas turbine technologies discussed in Chapter 8.

Investors may be nervous about the long-term reliability of the machines given the youth of the technology. Wind turbines are subject to complex, continually varying stresses. There have, consequently, been failures even in quite high quality machines, particularly with cracks in blade roots and primary shafts. Manufacturers are confident they have mastered the worst of the problems, and the availability of the best windfarms installed in the mid-1980s has exceeded 90%. Manufacturers usually offer insurance for the first few years (Mitsubishi offer 10 year cover), and longer-term insurance is available, reflecting in part the relative ease and limited costs of replacing even major components.

Wind energy has been deployed in a variety of configurations from arrays of hundreds of machines in California, to community-owned isolated turbines in Denmark. Unless local interests are prepared to manage isolated machines, there are strong incentives for installing arrays of sufficient size to distribute various fixed base and connection costs, and to employ maintenance personnel within easy access. In countries with lightly populated wind-swept expanses, large arrays in wind producing regions (like the Californian Altamont Pass) may emerge. In more populous regions such as western Europe which lack the open expanses of the US, windfarms of around 5-10MW, costing £3-6m and occupying about 1km^2, seem likely to dominate, though many machines may also be installed singly or in small clusters.

23. Because of exchange rate fluctuations, quoted costs are kept in £/p or $/¢ according to original source. In December 1991, £1=$1.85.

Table 10.3 US wind energy cost trends and projections, 1981-2000 (1990US$)

	1981-85	1986-91	1992-95	1996-2000	Post 2000
Installed cost, $/m^2	650	460	400	400	350
Availability (%)	60	90	90	95	95
O&M, ¢/kWh	2.5	1.5	1.1	0.7	0.5
Annual energy, kWh/m^2/yr:					
Good site, 350W/m^2	350	500	630	750	750
Excellent, 500W/m^2	500	750	1025	1,100	1,100
Energy cost,¢/kWh					
Interest rate 6%					
site 350W/m^2	18.3	9.5	6.7	5.4	4.7
site 500W/m^2	13.6	6.9	4.6	4.0	3.4
Interest rate 12%					
site 350W/m^2	27.4	14.0	9.8	8.1	7.0
site 500W/m^2	20.0	9.9	6.6	5.8	5.0

Assumptions: lifetime 25 years; land rent 0.3¢/kWh; annual insurance = capital x 0.5%. 'Excellent' sites: mean windspeed 7-8m/s at hub height (class five or higher in US wind atlas). 'Good' sites: mean windspeed 6-7m/s (mid-range in US wind atlas).

Data:1981-85: data from Altamont Pass, California.
1986-91: commercial data for US Windpower and Danish machines around 100kW
1992-95: cited performance of prototypes available in 1991
1996-2000: projections by US Solar Energy Research Institute
Post 2000: mass production estimates based on automobile weight costs

Source: D.Cavallo, D.Smith and S.Hock, 'Wind Energy: Technological and Economic Aspects', in T.B.Johansson et al, (eds), *Fuels and Electricity from Renewable Sources of Energy*, Island Press, Washington DC, forthcoming 1992.

Wind turbines can also be sited offshore. Problems of wave loading, salt erosion, access and transmission difficulties make it a more difficult and costly environment, but this is offset at least in part by the much stronger winds at sea. Some windfarms have already been installed on levees going into the sea, and in shallow protected waters. The first genuinely offshore station, albeit in relatively shallow waters of 2.5-5m of water, was installed in 1991 and is estimated to have generation costs about 40% above those of onshore systems.[24] Studies for deeper waters (above 5m) have tended to assume very large machines, because of the large fixed cost of foundations, with somewhat speculative (and probably dated) estimates of costs at 1-3 times that of onshore wind energy, depending both on assumptions and the nature of the location.

10.4 Resources and market potential

Estimating the potential contribution of wind energy to future energy supplies is much more complex than for most energy sources because resources and costs vary so widely, and in many cases are so poorly known.

A crude estimate of baseline technical potential, taking account of machine performance and interference, but neglecting economic and siting constraints, can be made without difficulty. Even in a relatively crowded region with high electricity demand, such as western Europe, the technical potential amounts to several times current electricity demand.[25] But the amount which can be exploited in reality depends upon a host of uncertain technical and economic assumptions, and above all upon the assumed availability of sites. This is not primarily a matter of land use (the land rendered unavailable for other uses is comparable to conventional sources[26]) but of visual impact.

Many people, pointing to the size of the larger wind turbines (typically 25-40m high, with blades of a similar diameter), and the intricacies of

24. 'Making History at Sea', *Windpower Monthly*, September 1991. The 11x450kW installation, off a Danish island, is stated to have cost about twice that of an onshore installation, but after allowing for the stronger offshore winds the energy costs are about 40% higher.

25. Typical resource densities are 100-300W/m^2 at hub height. With 30% conversion efficiency and machine spacing at 10 times the rotor diameter, a spacing at which the energy extracted is largely replaced from the upper atmosphere, simple calculation demonstrates the energy available to be several times the density of electricity demand.

26. Gipe, op.cit.

local planning procedures, believe that local objections to siting will limit the practical resource to a small fraction of national demand. Others express the view that wind turbines are greatly preferable to coal or nuclear generation, and can be made quite attractive. They believe that wind turbines could be placed at many sites, providing that basic first-order siting restrictions are met - no machines too close to houses or in national parks, none too near TV or radio transmitters, none in difficult or dangerous locations such as airports. The resource would then be very large. Even in some European countries (eg. the UK and Greece) the theoretical potential could be comparable to total electricity demand.[27] In more sparsely populated countries, such as the US, the potential would be greater still.[28]

There is little objective evidence on which to choose between these extremes. Objections in California and Europe have prevented some windfarms proceeding, but have not seriously impeded overall developments in most areas. Public opinion surveys, near windfarms or using pictures and videos, have often shown positive reactions; but noisy or badly designed windfarms have also raised strong opposition.[29]

---

27. An EEC study estimated the wind resource in the EC-10 (ie. 1990 membership less Spain and Portugal) after such first order siting restrictions and primary conversion losses to be 4,000TWh/yr, more than three times EEC electricity demand. Figures for the UK and Greece were 1,760TWh and 550TWh respectively, both many times national demand (H. Selzer, 'Wind resource assessment', *European Solar Energy R&D, Series G. Vol.2*, 1986). These estimates assumed very large machines and lenient first order siting restrictions.

28. 'Although this study shows that, after exclusions, only about 0.6% of the land area in the contiguous United States is characterised by high wind resource (comparable to that found in windy areas of California ...) the wind electricity potential that could be extracted from today's technology with these areas across the United States is equivalent to about 20% of the current US electric consumption ... as advances in turbine technology allow areas of moderate wind resource to be developed, more than a tenfold increase in the wind energy potential is possible ... to produce more than three times the nation's current electric consumption' (D.L.Elliot, L.L.Windell, G.L.Gower, 'An assessment of the available windy land area and wind energy potential in the contiguous United States', Pacific Northwest Laboratory, PNL-7789, August 1991).

29. I. Calman, 'The views of politicians and decision-makers on planning for the use of windpower in Sweden', *European Wind Energy Conference, Hamburg, 1984*, and 'Public opinion on the use of wind power in Sweden', *European Wind Energy Conference*, Rome, 1986. F. Lubbers, 'Research program concerning the social and environmental aspects related to the windfarm project of the Dutch electricity generating board', *CEC Wind Energy Conference*, Herning, 1988.

Experience in Denmark has indicated wide variations, in part depending on ownership: people have proved far more receptive to community-sponsored projects, with which they feel more involved and perceive direct benefits, than those installed directly by utilities. In the UK, where the economic conditions have forced developers to seek windy sites on high ground, environmental objections are slowing down initial development, though the industry is confident that once people become more familiar with wind turbines and less intrusive sites become economically feasible, objections will ease. Experience to date thus suggests that siting difficulties will slow the expansion of wind energy, but it is too early to say how serious the long-term constraints will be; the possibility of large-scale wind energy certainly cannot be ruled out.

Estimates of 'how much wind energy is practical?' are thus essentially subjective. A more useful question is 'what would be required to extract a certain amount of wind energy?' In Britain, a rough answer is that for each 1% of electricity demand supplied by wind energy, about one hundred, 10MW arrays would be required (each with at least 20 machines at current commercial sizes), sited on about 0.1% of mainland area (perhaps 150-250km^2) including site boundaries (as noted above, much of the occupied land could still be used for farming).[30]

A second tier of uncertainty arises from the variability of wind economics between sites and systems. The energy output depends heavily on the windspeed, which varies strongly with local conditions as well as with the region of deployment. The installation cost similarly varies according to local conditions for foundations, access and connection. The value of the electricity depends on the marginal costs of the local power system, which also vary widely.

But as a broad generalization, wind energy in most regions with mean windspeeds at 30m height above about 7m/s is already competitive unaided, and expected cost reductions over the next few years should make many regions above 6m/s of commercial interest, depending in part on gas availability and price. Figure 10.4 shows estimates of wind resources in Europe, which suggests that competitive areas include the west of Britain and Ireland, most of the north European coastline, and internal mountainous areas.[31] More generally, coastal regions within

30. M.J.Grubb, 'Wind Energy in Britain and Europe: How Much, How Fast?', *Energy Exploration and Exploitation*, Vol.8 No.3, 1990.
31. ibid.

Figure 10.4 Distribution of European wind energy

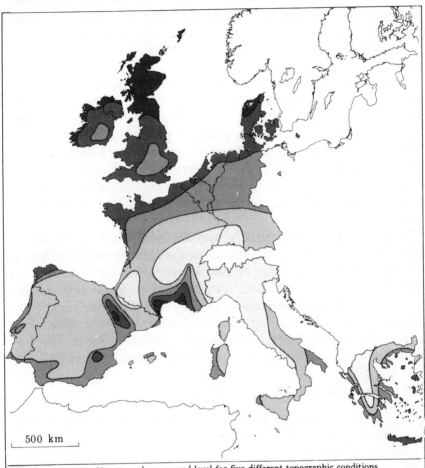

	Wind resources at 50 metres above ground level for five different topographic conditions									
	Sheltered terrain		Open plain		At a sea coast		Open sea		Hills and ridges	
	m s⁻¹	Wm⁻²	m s⁻¹	Wm⁻²	m s⁻¹	Wm⁻²	m s⁻¹	Wm⁻²	m s⁻¹	Wm⁻²
	> 6.0	> 250	> 7.5	> 500	> 8.5	> 700	> 9.0	> 800	> 11.5	> 1800
	5.0-6.0	150-250	6.5-7.5	300-500	7.0-8.5	400-700	8.0-9.0	600-800	10.0-11.5	1200-1800
	4.5-5.0	100-150	5.5-6.5	200-300	6.0-7.0	250-400	7.0-8.0	400-600	8.5-10.0	700-1200
	3.5-4.5	50-100	4.5-5.5	100-200	5.0-6.0	150-250	5.5-7.0	200-400	7.0-8.5	400-700
	< 3.5	< 50	< 4.5	< 100	< 5.0	< 150	< 5.5	< 200	< 7.0	< 400

500 km

Source: I.Troen and E.L.Peterson, European Wind Atlas, Riso National
Laboratory, Roskilde, Denmark, 1989.

temperate zones, areas within strong trade wind belts, and mountainous areas, are likely to provide economic wind energy. Exceptional thermal forcing effects, such as those responsible for the Californian wind resources, can also create currently economic resources. As wind technology develops further the areas in which it is competitive will spread.

Installed costs would have to halve at least before the bulk of low-windspeed sites would approach competitiveness. Again it should be emphasized that local factors can produce great variations about the regional average, and there are probably pockets of economic wind resources in most countries. Market expressions of environmental concerns, through constraints on conventional plant construction, incentives or quotas for non-fossil sources, or explicit pollution/resource taxes, will of course enhance the relative economic position of wind energy.

Concern over the constraints on onshore windpower has stimulated interest in siting offshore. As noted above the costs are uncertain, but not necessarily prohibitive. Siting in shallow waters could for some countries greatly increase the exploitable resource, and exploitation of broader areas of continental shelves could open up very large resources; in the UK, the potential for machines sited in 5m-30m of water, *after* taking account of shipping lanes, and fishing and military zones, has been estimated to be comparable with total UK electricity demand.[32]

10.5 Market obstacles and current trends

The apparent potential offered by wind energy contrasts forcibly with its negligible role in the current electricity market, and in future projections by mainstream energy business.

One reason for this contrast is simply the pace of wind energy developments over the last decade. The traditionally conservative utility business is used to technologies which measure in hundreds of megawatts and take a dozen years to plan and construct, let alone develop. The gulf between this and a technology which has passed through three complete phases of development and deployment within

32. D.Milborrow et al, 'The UK offshore windpower resource', *Proc 4th International Symposium on Wind Energy Systems*, Stockholm, BHRA, Cranfield, March 1992; K.Newton, *The UK Wind Energy Resource*, ETSU-R20, Energy Technology Support Unit, Harwell, September 1983.

this time, and has a typical unit size around a thousandth of the traditional scale, could hardly be greater. Management attention which is focused on developing conventional units of up to 1,000MW has little time for or interest in little-known technologies of such a small nature. Some utilities are simply unaware of the current state of the technology and find it hard to take seriously.

The sour taste left by some aspects of the US experience, already alluded to, is another major factor. This has been compounded by the fact that the incentives which led to the 'Californian windrush' were restricted to independent companies, selling electricity to utilities (which were forced to buy it). It was felt that monopoly utilities would be reluctant to pursue wind energy, but this became a self-fulfilling prophecy, inhibiting the development of utility expertise and labelling wind energy as an 'independents' activity.

Added to this is the complexity of wind energy as an investment option. Its performance and economics depend on a wide variety of siting conditions. Uncertainties over possible local impacts and a reluctance to accept that such variable sources can be as valuable as conventional sources are also important. These complexities, coupled with the speed of developments, mean that there are few people outside the wind industry itself with the expertise and confidence to judge investments.

The uncertainty is compounded by inadequate knowledge of wind energy resources in many areas. Wind resources are complex, and meteorological data on windspeeds is generally quite inadequate to identify areas and sites of real potential. Indeed, the wind speed maps for California available in the 1970s suggested that there were no areas suitable for the development of wind energy. The Californian Energy Commission Surveys, which identified the strong winds in the broad mountain pass areas, were a critical factor leading to the development of wind energy in California: one leading official commented that 'the wind [business] ... could have gone elsewhere in the US, but the data were just not there'.[33]

Electricity supply is a utility monopoly in most of the world. The trend towards increasing opportunities for private generation, discussed in Chapter 2, may help wind energy; but the parallel change in financial requirements discourages such capital intensive investments. The

---

33. Quoted in Ros Davidson, 'The California Covenant', *Windpower Monthly*, July 1989.

economics for independent generators depends critically upon the terms under which power can be sold to the grid; the cost paid for electricity has been the key focus, but another aspect is the terms for grid connection of windfarms which are not near existing power lines. Traditionally, utilities have funded extensions out of central funds, recouping costs over time as more users and/or generators connect to an extension. Loading the full costs of extension on to the first windfarm can readily make it uneconomic, an issue of growing concern.[34] Thus, utility liberalization carries mixed benefits, and even in the extreme case of privatization of the industry in the UK the real effect will depend heavily on the fine print of legislation and regulation.[35]

Combined with the changing attitudes of some utilities, the rise of environmental concerns, the increasing degree of difficulty in finding sites for conventional power stations, and continued development of the wind technology, wind energy markets will continue to expand. The combined targets of European governments total 4,000MW (4GW) of installed capacity by the year 2000; current trends, and the track record in meeting intermediate targets to date, suggest that this level will be reached, but perhaps slightly later than the target date. The European Wind Energy Association has presented a view of longer-term developments suggesting further targets of 10GW by 2010 and 100GW by 2030, which would equate to about 10% of current EC generation.[36] In the US, resumed activities in California may be joined or even outpaced by developments in the states dubbed the 'Saudi Arabias of wind', most notably Minnesota following the passage of a law aimed at opening up that State's vast resources.[37]

Developing country markets may also open up rapidly. India is moving quickly in the field and Denmark is supporting an Egyptian programme,[38] whilst China has a large programme for isolated

34. 'Utility monopoly causes market insecurity', *Windpower Monthly*, July 1991; 'Grid connection - a dangerous monopoly', *Windpower Monthly*, September 1991.
35. 'The Working Group do not believe that privatisation will work to the unmixed benefit of renewable resource development. New institutional and financial factors would seem positively to harm the prospects' (M.Laughton, (ed), *Renewable Energy Sources*, Watt Committee/Elsevier, London, 1990, p.9).
36. European Wind Energy Association, op.cit.
37. 'Minnesota wind power underway', *Windpower Monthly*, July 1991.
38. 'Going for full-scale manufacture in India', *Windpower Monthly*, January 1991; 'Danish aid launches Egypt project', *Windpower Monthly*, March 1991.

energy

supplies. But in general, the shortage of capital and other market factors will limit the rate of deployment unless imaginative - and perhaps risky - financing schemes are offered by the manufacturers or their governments. Finally, there will be a continuing and important market for island and other remote applications.[39]

10.6 Policy options

Governments and potential investors are faced with a common dilemma. Future developments will lower costs, giving a temptation to wait. Yet operating experience is a major need at present, and the development process itself will be hastened by large orders early on. Early experience is also required to help resolve various uncertainties about siting. The experience described above demonstrates that many diverse factors other than energy and technology prices alone affect wind energy development; governments have a determining role in removing some of the obstacles discussed above and more generally shaping the market for wind energy.

The fact that resource ignorance has constrained development highlights the importance of *wind energy resource surveys*. Fledgling companies cannot afford large-scale resource surveys, especially since if published the results would be as useful to competitors as to themselves. Efforts by the European Commission have led to the European Wind Atlas, and work by the Pacific Northwest Laboratory in the US has created a somewhat less detailed US atlas. These are important but still have significant limitations, and for local siting need to be backed up by further measurements; in other areas of the world, the data are much poorer still.

Electric utilities have not to date been in the forefront of encouraging wind energy for reasons noted above. In addition to greater enthusiasm, independent generators have some inherent advantages. Wind deployment requires knowledge of local conditions, not just of wind resources but of land uses, which may give advantages to locally-based windfarm companies. Furthermore the greater public acceptance of private generation (as noted for Denmark above) may be important. Yet independent producers cannot profit from wind energy without adequate payments for the electricity produced. *Utility liberalization and buy-back regulation*, which determines the conditions under which

39. 'Many and varied uses of wind power', *Windpower Monthly*, February 1991.

power may be produced and sold independently, is thus a central issue. Where independent generation is allowed, but with little oversight over the 'buy-back' tariffs paid by utilities for independently-produced power, independent generation has in practice been squeezed out (as with the UK Energy Act of 1983). A key point of the US legislation which helped to open the Californian market was the avenue it opened for regulation of such tariffs by utility commissions.[40] Legislation introduced in Germany in 1990 requires utilities to pay 90% of the delivered electricity price to independently generated power, and this has been largely responsible for the boom in planned wind power developments. The terms for grid connection, as noted above, can also be important.

Once these issues are addressed, it is *market stimulation incentives* which have made wind energy viable. The rationale for explicit support of new energy sources is simple. They face many obstacles, including simple barriers of scale and more direct manipulation of market power by established interests. Existing sources may themselves be subsidized. Most are considered to involve higher 'external costs' than wind energy, arising from their various environmental impacts and the implicit risks of dependence upon foreign sources. As noted in Chapter 2, the most widely-quoted study of external costs in Europe estimated them to be around 1.5-3.0p/kWh (0.02-0.04ECU/kWh) for many conventional sources, a level which the wind industry has been quick to point out would be sufficient to close the gap between wind and conventional power costs in many cases.[41]

If such externalities are not reflected in conventional energy prices, subsidizing alternatives is the obvious surrogate. The form of subsidy is important. Per-capacity subsidies encourage over-rating of machines (see footnote 21). Other forms of capital subsidy are possible, but in many ways a better approach is to support higher payments for the electricity produced. This directly rewards improved machine performance and can be justified on grounds of avoided external costs. However, unlike capital

---

40. In principle utilities were required to pay independent generators the 'full avoided cost', ie. the savings the utility made by not having to generate. In practice this could be interpreted in various ways, depending in part on estimates of capacity contributions and utility long-run marginal costs. Fixed price contracts were crucial to wind investments after the removal of tax credits, but have been another source of irritation for utilities forced to buy wind power even at times when the price greatly exceeds their own generating costs.
41. European Wind Energy Association, op.cit.

subsidies a key requirement for government support of this kind is some evidence that it will not suddenly be withdrawn: 'any incentive scheme must recognize the capital intensive nature of wind energy and hence provide a framework with a sufficiently long time span to permit commercial investment over the lifetime of a wind turbine'.[42] Confidence will be greatly increased if legislation contains commitments or recommendations as to duration of support. Also, market stimulation programmes can be directed towards quality goals - the Dutch programme for example includes bonuses for quiet wind turbines and those sited within designated areas.

Utility and national targets could be an alternative to subsidies if wind energy developments were led by utilities that were compelled to meet such targets, and may appear unnecessary if conditions make wind profitable for independent investors. In practice, they appear quite important as a complement to other measures. Realistic targets give companies an indication of how large the national market will be, and hence the extent of development investment which may be justified. They also provide a ready measure against which to judge actual progress, with an overall sense of coherence to wind energy prospects and development. There are however clearly limits to this; targets need to be flexible enough to accommodate developments, and cannot be set rigidly too far ahead.

National certification reduces the danger that by relying solely on market forces backed by strong incentives for a fledgling industry, people will seek a 'fast buck' with inadequate technology. Even if poor companies are weeded out over time, many resources may be devoted to poor machines, and the reputation of wind energy undoubtedly suffers. Licensing standards and national design testing and certification have helped to maintain quality in Denmark and the Netherlands, and could have avoided many of the worst developments in the US.

Suitable planning processes and designated areas would reduce planning delays and uncertainties. Inadequate planning can result in chaotic and unsightly developments, and mounting public opposition. The converse can be equally damaging, if every windfarm has to go through lengthy and costly procedures through which a single objector can effectively block developments. Some countries have specified 'designated areas' for wind development in favourable areas, in which

---

42. ibid.

streamlined planning procedures ease development. However, a great deal of the responsibility will always rest with wind developers, to develop designs and sites with sensitivity and sympathy.

Developing countries offer great scope for wind energy, but capital constraints could greatly inhibit wind's development in poor countries. Wind energy could be a major element for funding in *international technology transfer* arrangements for meeting global environmental concerns. More generally, *capital support for wind exports to poor countries* from donor countries or companies, and some funding for *joint ventures* to help establish indigenous capabilities could greatly accelerate the development of wind energy in poorer countries.

Finally, *research and development* remains important. The wind industry itself funds steady incremental improvements. In Europe developments have been considerably aided by seed money from the European Commission to assist private R&D, and for 'common good' technical and resource research. The collaboration of the US Windpower Company with utility finance to develop the variable-speed machine cited above marks a new departure, but substantial government money is still likely to be important in funding major new avenues, such as dedicated offshore machines. Yet one lesson of the past fifteen years is that government projects alone have yielded limited benefits; the large strides have been made through public support for, or collaboration with, private endeavours.

10.7 Conclusions

Wind energy has developed very rapidly during the 1980s. The economics vary greatly according to the siting and financial conditions, but it is already competitive without support in the most favourable conditions, and modest levels of government support have opened up rapidly growing markets. Continuing developments will enhance its position. Current trends suggest a market during the 1990s focused on Europe and the US, with several thousand megawatts installed in each; developing country plans, dominated by India, comprise a similar magnitude.

The rate of growth will however depend strongly on the scale, nature and consistency of government policies. The most important factors are:

* policies which affect the value of electricity generated by wind energy both within utilities and by independent generators (including utility financing conditions, the degree of subsidies to existing sources and/or internalization of external costs, and conditions for independent power production);

* direct support for wind energy to overcome initial hurdles and reflect the environmental benefits over other options;

* protection of independent wind generation against utility monopoly powers;

* streamlining of planning procedures; and

* wind resource surveys.

Also, support for transferring the technologies to developing countries to overcome capital constraints would be required for rapid growth in these regions.

Wind energy resources vary considerably but in many regions are large. The industry is still in its early days and many factors remain uncertain, but wind energy has clear potential to become an important component in power generation during the first few decades of the next century.

Solar Electricity from Photovoltaics

Photovoltaic (PV) generation of electricity, is making an increasing contribution to the supply of energy services, at present mainly in countries with high levels of sunlight. PV can provide autonomous power supplies in locations not connected to a grid system more cheaply than other technologies, even in temperate climates. PV is an economic alternative now to conventional peak power plant in areas with daytime peak loads and abundant sunshine. PV systems used as embedded generation on long distribution lines can greatly improve voltage regulation power factors and harmonic content and thus have significant value for utilities. Ground-based PV systems are being installed and studied in a number of European countries and the use of PV cladding on buildings is being actively pursued, the latter being seen as a promising prospect for the UK. Industry and academia continue to take a leading role in R&D in the UK, but government and international core support may prove crucial if the UK PV industry is to develop as rapidly as possible.

The major role of PV in developing countries is to provide the power for basic necessities such as light, clean water, communications and cold storage for food and medical supplies. Rural electrification using PV systems, usually small systems on each house, is proceeding rapidly in a number of countries. A purely commercial market for PV lighting, battery charging, TV or radio power etc is growing in many developing countries. The major constraints are the financing of the initial capital cost and the lack of technical and distribution infrastructure. Programmes to enable local financial bodies to offer loans or hire purchase, and training programmes for local companies are addressing these needs with great success. The utilities in developing countries could also improve the reliability of supply quite significantly by the use of embedded PV generation.

PV costs are expected to continue to fall because of both increased scale of production and the introduction of technical advances, thus making a wider range of applications economic in temperate climates as well as in areas of high insolation. PV can make an important contribution to the social and economic advancement of developing countries and can also make an environmentally benign contribution to electricity generation worldwide from the late 1990s onwards.

11.1 Characteristics of the technology

Solar cells convert light into electricity. Solar cells are basically large area semiconductor junction devices, and the light is converted into electricity within these junctions by the 'photovoltaic effect'. The whole technology of converting light to electricity and using that power to perform some useful service is often known as photovoltaics or PV for short.

A solar cell produces direct current (DC) electrical output, with the current being proportional to both the area of the cell and the intensity of the light. The voltage depends on the type of semiconductor used to make the cell and there is also a weak (logarithmic) dependence on light intensity.

The output of a solar cell is usually specified in terms of peak Watts (Wp) or peak current, which is the output under the internationally agreed standard conditions of $1kW/m^2$ sunlight intensity (almost the maximum level achieved under bright direct sunlight) and a cell temperature of $25^{\circ}C$. One 10cm x 10cm solar cell made from crystalline silicon would give an output under those conditions of 1.25-1.5Wp (resulting from a current of about 25-30milliAmps/cm^2 at a voltage of $^1/_2$ Volt).

Solar cells must be protected against breakage and corrosion if they are to achieve a 20-30 year lifetime. They are usually incorporated in a *module* which is a string of cells connected in series and encapsulated in glass or plastic. The series connected cells give an output equal to the current of each cell, and the sum of their voltage. Usually 30-36 cells such as those referred to above are connected in series to give an output of more than 12V even on less sunny days so that a 12V battery can be charged.

These modules are the standard product of the PV industry, but their 35-50Wp output is too small for many purposes. To obtain higher power, the modules are connected together in an *array*. An array is a collection of modules connected together to give the required current and voltage output and held in a frame which must be strong enough to resist wind loads. The modules are orientated in the array to collect maximum sunlight. In a stationary array this usually means that the modules face the equator and are tilted at an angle to the horizontal about equal to the angle of latitude. Movable arrays track the sun as it moves from east to west during the day (1 axis tracking) or are fully steerable to point always directly at the sun throughout the year (2 axis tracking).

PV modules convert sunlight into electricity, but this electricity is rarely useful to society unless it is, in its turn, converted to some service such as lighting or motive power etc. The cost of the service depends on the cost and efficiency of the entire PV system, so the costs additional to those of the modules - called the balance of system (BOS) costs - must also be minimized. The BOS costs have two components, those which relate to the area of the system such as land costs and array construction costs, and those which relate to the power, such as the costs of electronic controls, storage batteries, DC to alternating current (AC) inverters and loads such as water pumps or lights. The area BOS costs are inversely related to module efficiency, and reducing these costs is one of the driving forces in the search for higher cell and module efficiencies.

The efficiency and reliability of the system components downstream of the modules is crucial to the performance of the whole system, and these components have often been less reliable than the PV modules. Batteries, in particular, are expensive, have a lifetime of only 5-7 years even with careful treatment and can form the weakest link in the system. The electronic control equipment and the inverters have in the past also been the cause of unreliability, but modern designs have proved to be both reliable and efficient.

11.2 Resource characteristics and environmental impact

Sunlight is a diffuse energy source with a maximum energy density of about $1kW/m^2$ at the earth's surface. It is also variable on timescales from seconds to decades, but particularly from day to night. In total however the resource is immense; the earth intercepts solar energy at over 10,000 times the rate of human energy consumption. Even after allowing for considerable conversion losses, the solar energy theoretically available from photovoltaic conversion in most countries greatly exceeds energy consumption.

The energy available varies greatly with location. In sub-tropical areas, the average insolation is frequently about half that at the peak time on a clear day. Taking into account cloud cover and seasonal variations, the annual average insolation for sites within about 35 degrees of the equator is typically about $1,600\text{-}2,000kWh/m^2$ per annum, about $200W/m^2$ (average power), though it can range up to $2,500kWh/m^2$ per annum for sites such as the Texas deserts and down to $1,100kWh/m^2$ per annum for cloudy areas such as Tokyo. The average falls off rapidly at higher latitudes especially for winter months; at $60°N$, the average daily insolation in winter is only 8% of the peak even on a clear day. The 'load factor' of PV plants - the ratio of average output to the peak - varies similarly. However, the average annual insolation in mid-latitude regions such as southern England is still almost half that of typical desert areas.

The variability of the input solar energy has a major impact on the applications of photovoltaics and its cost effectiveness. The ideal application for power generation is one where the demand matches the solar energy availability, as occurs for example with the peaking loads of utilities in the southern US, where air-conditioning is the major factor, but many other applications are of interest as discussed below.

The low energy density of sunlight requires large areas of PV modules if high power is to be generated. But whilst 100MWp requires about $1km^2$ of module area, only a few hundred square meters of land are necessarily rendered unavailable for other uses such as farming. In fact the land area rendered unavailable for other uses by PV generators has been shown by the US Department of Energy (DOE) to be about the same as for coal or nuclear plant of the same annual energy output when all factors are considered.[1]

1. US DOE Report, 'Energy Systems Emissions and Materials Requirements', US Department of Energy, Washington DC, 1989.

Apart from land impacts, photovoltaics is almost completely benign in operation. There are small occupational hazards in the maintenance of the systems, similar to those of a window cleaner, but most systems are low voltage and hence without serious electrical risks.

The production of PV systems is an industrial process and as such creates both environmental impacts and occupational hazards. The largest mass of a system is of glass, but the majority of the hazards are associated with the semiconductor materials. As a reasonable approximation, the environmental impacts and occupational hazards of PV systems are similar to those of the same value of products of the semiconductor industry. In a good solar climate, the hazards per annual GWh generated are lower than those of conventional utility generating stations.[2] Some of the materials of cell construction are potentially hazardous, so care needs to be taken in the manufacture and disposal of redundant cells.

11.3 Applications

At present prices, power from photovoltaics is much more expensive than that from conventional bulk electricity sources. Nevertheless there are widespread applications arising from the already distributed solar input and the simple and modular characteristics of PV, and other markets are expanding as costs continue to fall.

Most people first meet PV in the ubiquitous solar calculator of which 100 million per year are sold. Such consumer products form a significant market for PV manufacturers with beneficial effects on their cash flow, scale of output and funds for R&D. For instance one million 1 Watt garden lights sell per year, and this 1MW output is 2% of total world PV production. It is now cheaper to install a PV lighting unit in a shed at the bottom of the garden in the UK rather than to run a mains cable to the shed. For sailing boats and caravans etc PV is popular as a battery charger. Numerous consumer and leisure products are now switching to PV power, and these markets could take tens of megawatts per year in the 1990s.[3]

Another specialized market is for 'professional systems' where reliability is paramount and PV usually acts as small autonomous

2. US DOE, op.cit.

3. R.Hill and J.Day, Seminar on PV Markets, *Proc. 21st IEEE Photovoltaics Specialists Conference*, 1990, Institution of Electrical and Electronic Engineers, New York, 1990.

power-sources for equipment. Communications repeater stations and remote monitoring stations are now very commonly PV powered, and cathodic protection of pipelines, bridges and similar structures is usually most cost-effectively carried out by PV. Motorway signs and telephone points, even in the UK, can be more cheaply powered by PV than via long cable runs, and there is a rapidly growing market for military field applications.

The alternative to grid-connected systems for large power supplies is the 'stand-alone' system, where PV systems, sometimes in combination with other types of generator, form an autonomous power source which is not connected to a grid. A constraint on the use of all-PV stand-alone systems is the need to store electricity. Secondary batteries have improved in the past 10 years, but major improvement in the cost of storage, factors of 10 or more, will require new technologies.

Without such new storage technologies, large stand-alone PV systems are likely to be hybrids, combining PV with diesel or hydro-generators. In Australia, with its restricted grid system, a remote area power system (RAPS) has been developed. RAPS are typically of a few kilowatts size and consist of a PV array, battery storage, inverter and a diesel back-up generator. They supply power more cheaply than any alternative in the outback, and the Australian government provides a 25% rebate on the capital cost and a 50% subsidy on maintenance costs as a measure of equity between rural and urban areas. Much larger systems, up to 200kWp, will cost-effectively complement diesel generators for remote local grids when PV module costs fall below US$3/Wp, expected within the early 1990s.[4]

As noted above, a particularly attractive application on grids is where peak electricity demand coincides with peak solar input, usually because of air conditions. In California, where there is a total of about 10MWp of grid-connected PV installed, the utility Pacific Gas and Electric has reported very favourably on the technical and economic performance. Other multi-megawatt plants are being constructed or planned in the US and with carefully designed financing structures, the cost of electricity produced by this new generation of PV plant will be very close to the

---

4. M.A.Green, *Solar Cells*, Vol.26, Nos.1&2, February 1989, pp.1-12.

US¢20/kWh cost of peak period electricity from present fossil fuel generators.[5]

PV systems have also been shown to be cost-effective now in providing daytime demand when installed at critical points in a grid distribution network.[6] As 'embedded generation' in this way the PV system avoids the cost of upgrading the lines and/or transformers at these critical and potentially overloaded points. This application is also likely to be highly cost-effective in the distributive systems of many developing countries, where it could avoid the brown-outs (excessive voltage reductions) or black-outs which are such an economically disruptive feature of these countries. For all utilities with long distribution lines, the use of PV embedded generation can improve voltage stability power factors and harmonic content, and can thus have an economic value to the utility far in excess of the value of the power output of the PV plant.

The seasonal variability of solar radiation increases the PV system costs in high latitude areas, since the systems must be sized to deliver the required output under the weak winter radiation. PV is most cost-effective in the 'sun-belt' from about 40°N to 40°S. This includes almost all of the developing countries, as well as southern USA and Australia.

The use of PV as a cladding on the walls or roofs of buildings avoids the land costs of ground-based systems and also avoids the cost of conventional building claddings. With PV modules replacing claddings costing from £50-500/m^2, this avoided cost greatly improves the economics of the PV system.[7] Germany has a '1,750 Roof' programme to install 2-4kWp on the roof of each of the 1,750 domestic houses. The Netherlands is also promoting the use of PV roofs, whilst the recent (May 1991) UK programme concentrates on PV cladding for the walls of commercial buildings. The vertical orientation of PV on south-facing walls maximizes the output from winter sunshine and minimizes the variation in output between summer and winter. The daytime load of commercial buildings also gives a better match to the supply from the

5. E.A.DeMeo, *Proc. Euroforum - New Energies Congress*, Saarbrucken, 24-28 October 1988, p.52-58.

6. D.S.Shugar, *Proc. 21st IEEE Photovoltaics Specialists Conference*, op.cit., pp.836-43.

7. W.H.Bloss et al, *Proc. 10th EC Photovoltaic Solar Energy Conference*, Lisbon, 1991, Kluwer Academic Publishers, 1991, pp.1295-1300.

PV cladding. In more northerly climes the building-integrated PV systems are likely to be the earliest cost-effective application for grid-connected PV systems.

The major role of photovoltaics in developing countries is as a tool for social and economic development, rather than as a major power producer. At least 10% of those villages yet to be electrified could most cost-effectively be powered by photovoltaics and those systems which have been installed have proved to be reliable and popular with the villagers. Where TV has been provided there is evidence of a fall in birthrate. The provision of lighting and clean drinking water is a major social benefit, and PV systems have been proven throughout the world to be more cost-effective than small diesels. Vaccine refrigeration by PV systems is cheaper than by kerosene refrigerators, by about 40% per potent vaccine dose, and has a reliability over 95% compared to about 60% for kerosene fridges. The reliability of PV power for telecommunications systems is typically 95-99%, so much higher than a diesel-powered system that usage per telephone rises considerably. Potential users may have to walk some kilometres to the telephone point, and are willing to make the journey only if they are confident of finding a working telephone.[8]

The power supplied to rural areas of developing countries may be only 10-20 Watts per person, but the services provided by these few Watts - light, clean water, potent vaccines and communications (telephone, radio or TV), have a great social benefit, far larger than may be suggested by their small contribution to the nett generating capacity of the country.

11.4 Market size and development

As illustrated in Figure 11.1, the global market for PV has expanded rapidly since the mid-1970s, with a sudden rise in 1983 due to large US government orders. The structure of the market has changed rapidly during the 1980s, as illustrated in Figure 11.2; US government orders declined sharply after 1983 but this was more than offset by increased commercial orders, which by 1987 accounted for the bulk of sales. The commercial and consumer markets are expanding at the rate of about 20-30% per year worldwide; total sales reached 46MWp in 1990 and are expected to exceed 50MWp in 1991. Taking all system components,

8. R.Hill, *Proc. 9th European Photovoltaics Solar Energy Conference*, September 25-29 1989, Freiburg, Kluwer Academic Publishers, 1991.

Figure 11.1 Annual global PV module production (MWp/year)

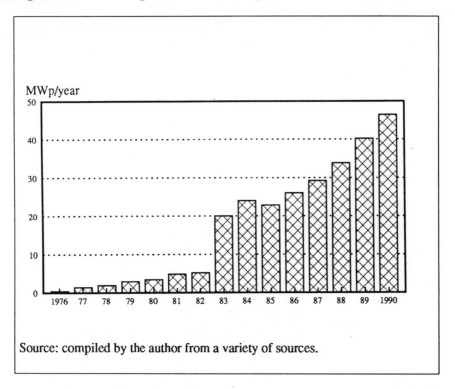

Source: compiled by the author from a variety of sources.

consultancy and design etc, as well as the PV hardware, the PV industry turnover worldwide in 1990 was in the order of £500 million.[9]

Interest in grid-connected PV is currently greatest in the southern US, where orders for PV as a peaking plant are expected to grow rapidly through the 1990s. The US DOE proposes an installed capacity 1.4GWp in the USA by the end of the century.[10] Grid-connected systems are also being installed in Europe, including Spain, Italy, Germany and Switzerland, with the latter country making a policy decision to install around 200MWp by the end of the decade.

The market for PV worldwide will depend strongly on cost reductions. Some published projections of the likely prices and markets in the next

9. P.D.Maycock, *Proc. 10th EC Photovoltaic Solar Energy Conference*, op.cit., pp.1396-1400.
10. R.H.Annan, *Proc 21st IEEE Photovoltaics Specialists Conference*, op.cit.

Figure 11.2 Market share of PV sales, 1983-87 (MW)

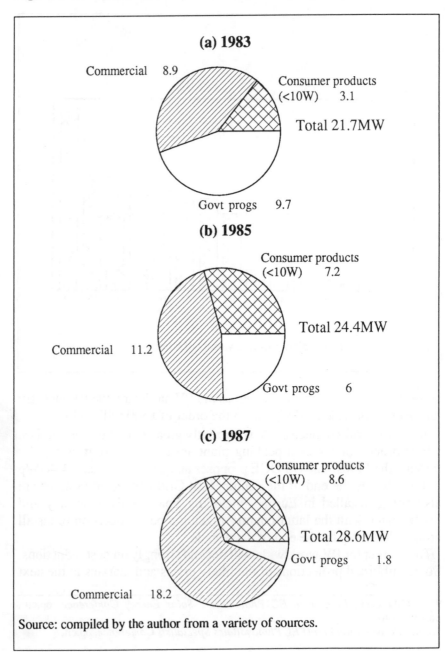

(a) 1983

Commercial   8.9

Consumer products (<10W)   3.1

Total 21.7MW

Govt progs   9.7

(b) 1985

Consumer products (<10W)   7.2

Total 24.4MW

Commercial   11.2

Govt progs   6

(c) 1987

Consumer products (<10W)   8.6

Total 28.6MW

Govt progs   1.8

Commercial   18.2

Source: compiled by the author from a variety of sources.

Table 11.1 Projection of PV prices and sales 1990-2000 (1991US$)

Year	Module cost ($/Wp)	System cost ($/Wp)	Sales (MWp/year)
1990	4.0	8-15	45
1995	3.0	6-10	100-200
2000	2.0	3-7	500-1000

Source: data compiled by the author from many sources, eg. Starr in W.H.Bloss et al, *Proc. 10th EC Photovoltaic Solar Energy Conference*, Lisbon, 1991, to be published. See also R.Hill and J.Day Seminar on PV Markets, *Proc. 21st IEEE Photovoltaics Specialists Conference, 1990*, Institute of Electrical and Electronic Engineers, New York, 1990.

10 years are summarized in Table 11.1. If cost projections are realized, sales could approach 1,000MWp annually by the end of the decade. The industry is confident that such targets will be attained, because of the many options for continuing development as discussed in the next section.

11.5 Photovoltaic technologies: past, present and future

The photovoltaic effect was discovered in 1839 by Becquerel in the same classic experiment which laid the foundations of modern photography. Although PV devices using selenium and copper oxide were commercially available in the 1920s and 30s, conversion efficiencies over 1% were not achieved until the discovery of the silicon p-n homojunction cell in 1954. The use of these cells for space satellites provided the technical impetus and the market pull for the development of PV cell technology through the 1960s, until the oil crisis of 1973 led governments to invest in ambient energy technologies. Major development programmes were initiated in the mid-1970s in the USA, Japan and Europe. The results of these programmes are evident in the present state of the PV industry, and in the range of technological options

Figure 11.3 History of solar cell efficiencies

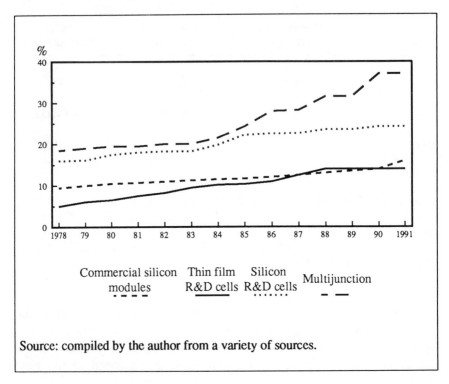

Source: compiled by the author from a variety of sources.

available, the most important of which are outlined in the box on page 209.

Figure 11.3 shows the increase in solar cell efficiency in the past ten years, for crystalline and thin film silicon, and for cells under concentrated sunlight. Module efficiencies have risen over that time from about 8% to 14-16% for standard commercial products using crystalline silicon, whilst PV array system efficiencies have increased from about 5% to 10-12%.

Figure 11.4 shows the module cost for purchases over about 1kWp. For major impact on the US southern electricity utilities, the target costs have been identified by the US DOE as shown in Table 11.2. Table 11.3 shows the materials and combinations which could meet these efficiency and cost goals, all of which are the subject of active R&D programmes. Module production costs of around £50/m^2 in full operation are

Figure 11.4 History of PV module costs

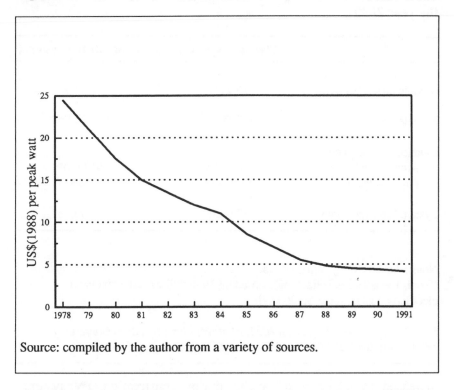

Source: compiled by the author from a variety of sources.

calculated for the 10MWp per year production facility for amorphous silicon in the USA.[11]

The prospects for the near term (early to mid-1990s) are shown in Table 11.4 for those materials which are in commercial production or pre-commercial development in 1991.

11.6 Policy issues in the application of photovoltaics technology

A key issue at the present time in the intercomparison of energy technologies is the basis on which these comparisons are made. The cost of energy is only partly determined by the technological success of particular processes - social, political and financial factors play an important part. The cost of energy from a capital intensive technology with a long construction time, such as nuclear or tidal, is substantially

11. Annan, *Solar Cells*, op.cit., p.135-48.

Table 11.2 Technical PV goals of the US government and industry for the year 2000

	Flat plate systems	Concentrator systems
Module efficiency at 25°C	15-20%	25-30%
Module cost	$45-80/m^2	$60-100/m^2
Balance of system costs Area-related[*] Power-related	$50-100/m^2 $150/kW	$125/m^2 $150/kW
System life expectancy	30 years	30 years

Notes: [*] balance of system costs vary depending on the type of flat plate system (fixed, 1 axis or 2 axis tracking). Based on 1986 dollars and a levelized electricity cost target of $0.06/kWh.

Source: R.H.Annan, *Proc. 21st IEEE Photovoltaics Specialists Conference, 1990*, Institute of Electrical and Electronic Engineers, New York, 1990.

increased by the interest payable during construction. PV systems, although they are also capital intensive, have short times between the start of planning and the earning of revenue and so minimize interest payable during construction.

In the UK, government interest in PV has been at a relatively low level, since the technology has been regarded as speculative. Now, building-integrated PV is seen as promising and warranting further effort to resolve uncertainties, particularly with regard to longer-term economics. The support of government and the new regional electricity companies is crucial to the success of the new national programme, in which industry and academia are playing the leading roles.

Another important issue is the question of social and environmental costs, or external costs. As noted in Chapter 2, inclusion of external costs would greatly improve the relative economics of PV.

Table 11.3 Research paths in the photovoltaic programme

	Flat plate systems		Concentrator systems	
	Single- junction	Multi- junction	Single- junction	Multi- junction
Amorphous thin films	a-Si:H a-Si:H:F	combined with: a-SiGe:H a-SiGe:H:F a-SiC:H a-SiC:H:F		
Polycrystalline thin films	CuInSe$_2$ CdTe Cu(In,Ga)(Se,S)$_2$ Cd(Zn,Hg,Mn) Te	combined with: HgCdTe HgZnTe		
Crystalline thin films	Silicon GaAs	GaAs/Si GaAs/Ge/Si	Silicon GaAs	AlGaAs/- -InGaAs GaAsP/- -GaAsSb GaAsP/Si GaAs/Si
Hybrid combinations(s)		CuInSe$_2$/a-Si:H CuInSe$_2$/GaAs		
Efficiency goals	10-15%	15-20%	20-30%	30-40%

Note: standard chemical nomenclature is used to describe the classes of compounds which form the PV cells. The most important are Si, silicon; CuInSe$_2$, is copper indium diselenide; CdTe, cadmium telluride, and GaAs is gallium arsenide. These are also referred to in Table 11.4. Other compounds are also being researched, as shown above. Other elements in other compounds are H, hydrogen; F, fluorine; Ge, germanium; C, carbon; P, phosporus; Sb, antimony. This is an indication of the complexity of the research programme.

Source:R.H.Annan, *Proc. 21st IEEE Photovoltaics Specialists Conference, 1990*, Institute of Electrical and Electronic Engineers, New York, 1990.

Table 11.4 Materials and efficiencies of PV devices available through the 1990s, with module costs estimated for different annual production rates

Cell material	ηC	ηM	MWp/yr	Module costs (£/Wp 1989) 10 MWp/yr	100 MWp/yr
Silicon (Si)					
Single crystal	24	14	3	1.5	1.0
Polycrystalline	20	12	3	1.2	0.8
Amorphous	14	4	2	1	0.4
Copper indium diselenide (CuInSe2)					
	14	>10	2	1	0.4
Cadmium telluride (CdTe)					
	14	>10	2	1	0.4
Concentrator cells					
Silicon (Si)	27	15-19	4	1.5	0.8
Gallium arsenide (GaAs)					
	29	-	3	1.2	0.7
Multijunction	37	-	4	1.0	0.5

ηC = best cell efficiency achieved in research laboratories
ηM = standard commercial module efficiencies
Note: for an explanation of the different PV technologies, see box facing.

Source: the estimates of production costs are for quantity purchase. The data for the commercially available modules are compiled by the author from a variety of sources. The data for CIS and CdTe are from the author's own calculations for the European Commission. The estimates for concentrator cells are based on US DOE data.

Characteristics of Photovoltaic Technologies

A number of different photovoltaic technologies are possible, which vary in characteristics such as cost and efficiency (see Table 11.2), stability and the stage of development.

Single-crystal silicon cells, in which silicon wafers are cut from a single cylindrical crystal, are a well established technology. These tend to be relative efficient and stable, but costly. *Polycrystalline cells* consist of silicon wafers cut from a cube of multiple crystals of silicon, each crystal being 1-10mm in size. This is easier to produce than single-crystal cells and results in square cells which fill space better, but they are less efficient than cells from a single crystal.

In *thin-film* cells, in contrast, the material is deposited directly on to the cell backing (substrate) in a non-crystalline form. This enables a layer as little as one-hundredth the thickness of crystalline cells to be used, deposited along with cells and interconnects in one continuous process. The much-reduced thickness enables more expensive materials to be considered.

Three major thin-film options are: *Amorphous silicon* cells are commercially available. The initial efficiency of a-Si modules is about 6-8% and they degrade over a year or so to about 3.5-4.0% where they are stable for years. *Copper indium diselenide* is at a pre-commercial stage, with a commercial product expected in shortly from one or more of three manufacturers. The stability appears to be excellent and 30cmx30cm modules with over 11% efficiency have been produced. *Cadmium telluride* is also at pre-commercial development stage, with commercially available modules expected in 1993. It shows excellent stability and module efficiencies over 10%.

Different materials can be stacked together to make *multi-junction* cells with efficiencies higher than is possible with any of the single materials but with higher costs; reducing costs and improving interfacing in such layers is an important research area.

Concentrator modules provide another way of reducing the requirements for the PV material. These use lenses or mirrors to concentrate sunlight, typically by a factor of 50-100, on to a small solar cell. Only silicon concentrators are commercially available in quantity at present. These cells need 2-axis tracking to ensure that the concentrated sunlight always falls on the cell, and will collect only the direct beam component of sunlight.

The market for grid-connected PV systems will be realized, even if cost and efficiency goals are met, only if other utility criteria are also met. The US DOE has published an assessment of these criteria, which shows that PV is potentially very attractive for utilities, particularly where environmental concern is strong and construction permits are difficult to acquire for other technologies.

In developing countries, PV is mainly in competition with diesel generators. Apart from the issue of government subsidies to diesel fuel, the main concern is the availability of capital. PV is already proven to provide electricity or energy services at lower cost than diesel on a life-cycle costing basis. Few individuals, however, use life-cycle costs as a basis for decisions on purchasing, and the initial capital outlay is their main criterion. The macroeconomic needs of the national economy to have the lowest cost energy options are thus confounded by the microeconomic perceptions of the purchasers of diesels with low capital cost but higher life-cycle costs. There is an urgent need to construct financial packages which convert the high capital cost of PV into a recurrent cost for the purchasers. The development banks themselves need soft loans and guarantees, provided as part of the aid from industrialized to developing countries. Initiatives such as the Global Environment Facility of the World Bank and the UN FINESSE programme are addressing these issues.

The UK could continue to play a leading role in PV applications for developing countries, but only if the nationally and internationally supported programme continues to provide core funding.

The other major initiative needed to enable developing countries to enjoy the benefits of PV concerns the provision of training and education. The ultimate aim must be the development of an endogenous capability within the developing countries, either on a national or regional level. There is a need for training of indigenous personnel at all levels and the establishment of centres of expertise from which the endogenous capabilities can develop. These needs are now clearly perceived by UN agencies and some developing countries, and steps are being taken to meet these needs.[12]

12. Proc UNCSTD meeting, Sao Paulo, Brazil, September 1991, forthcoming, UN, New York, 1992.

11.7 Future impact of photovoltaics

PV is a very young technology starting essentially in the mid-1970s. Since that time, the cost of PV modules has decreased by a factor of about 10, system efficiency has about doubled and module lifetimes have increased from 6 months to over 20 years. This rapid progress is continuing with new and improved products at all stages of development from speculative research to commercial development. The routes to module costs of £50/m^2 or less are proven, as are the routes to 30% efficient cells. The materials and technologies which will give the optimum combination of efficiency and cost are the subject of active R&D and of debate at present. It does however seem likely that this optimum will be different for the various PV markets. Power modules will need lifetimes of 30 years, efficiencies over 20% and costs of about £50/m^2, if PV power stations are to be competitive for utilities in the USA, Australia, Japan, southern Europe and elsewhere with high insolation levels. From the data presented in this chapter, it can be predicted with a high degree of confidence that such modules will be produced commercially early in the next century, selling in GWp quantities.

It is not yet proven that PV integrated into buildings will make a substantial contribution to electricity supply in higher latitude countries such as the UK. However, the results of studies now in progress will assess the technical contribution, and in addition will make economic projections based on several different PV module and fuel price scenarios. Preliminary results from the Netherlands and Germany suggest that this application could be commonplace from early in the next century.

The impact of PV in developing countries is potentially dramatic in terms of an improved quality of life for the rural poor. The major aid providers now see PV as a reliable and cost-effective technology and aid programmes are expanding considerably. Education, training, creation of centres of indigenous expertise and the development of endogenous capabilities will promote not only the application of PV, but expand the technology base of their society in ways which promote economic and social development.[13]

13. S.C.Trindade, *Proc. Euroforum - New Energies Congress*, Saarbrucken, 24-28 October, 1988, p.79-82.

11.8 Conclusions

Photovoltaics is environmentally benign in operation, its modular nature allows rapid construction and easy expansion and its energy source is distributed, free and reliable in the areas where the majority of the world's population live. The technology is maturing rapidly and the chain of innovations and developments now in place will ensure that the efficiency and cost targets are met within the next ten years or so.

Extensive markets for special applications and stand-alone systems in sunny areas are assured. Technology developments are also likely to bring PV within reach as a major component of grid systems in tropical and other very sunny areas, particularly for meeting daytime peaking demand and for broader grid applications. PV could also contribute to supplies in temperature climates such as the UK if module costs come down sufficiently, with the most promising applications being for systems integrated into buildings.

Such extensive developments will however be significantly accelerated by increased support for already active research and development programmes in industry and academia, and by energy prices which reflect external costs and utility reforms which encourage small-scale power generation. Given such developments, PV is set to be a major source of power for much of the world's population in the 21st Century, making a significant contribution to the mix of energy sources which will secure a sustainable future for mankind.

PART IV. SYNTHESIS

Case Study Comparison and Impacts

Eight technologies have been studied for their potential impact on energy supply and demand, and the policy issues which surround them. They range from those such as more efficient domestic appliances and combined cycle gas turbines which are already commercially available and making an impact on energy markets, to those whose impact currently lies in special markets, but for which the promise of continuing rapid development may during the 1990s bring them across the commercial threshold for large-scale energy system applications, as exemplified by photovoltaics.

This final part of the book examines the lessons which can be drawn from these case studies, and some of the broader implications for energy developments over the next decade and beyond. This chapter starts by reviewing the case studies with respect to the common themes which they reveal, and the important points of difference; and then proceeds to consider the potential impacts which these technologies might have on energy supply and demand in the coming decades. The following chapter then examines more closely some of the future pressures on energy systems as they may impact on emerging technologies.

12.1 Common themes

A few themes run through all the chapters. Perhaps the broadest is simply the sense of potential. Despite a century of continuous development of energy systems in industrial societies, and a particularly rapid expansion of 'modern' systems since World War II, there is no indication that energy technology development is stagnating. This study has moreover taken only a very limited sample of the full possibilities listed in Chapter 3.

Whatever problems the energy world faces, it is not short of ideas and technologies for helping to alleviate them.

A second feature is that most of the technologies examined have arisen and will progress in response to external needs and conditions. To some extent, several of the demand-side technologies have arisen from a process of autonomous 'technology push', in which manufacturing and process advances have occurred within industry prior to any specific demands. But even these have arisen in part from manufacturers' initiatives taken in response to the oil price shocks, while the rate and extent to which they are utilized will depend strongly upon changes in external factors. The others display primarily the response of technology to changing conditions - 'demand pull' - notably concerns about oil dependence and price fluctuations, environmental impacts, and increased availability of gas. The studies tend to reinforce the faith that when and where there is a will, technology has a way.

Of the various trends discussed in Chapter 2, a theme in all chapters is that of environmental pressures. These range from the very local (as in building energy management, where the benefits to the indoor working environment are highlighted as being as important as reduced energy consumption, and perhaps a stronger selling point) to broader urban, regional and global environmental factors. In energy saving, the benefits which a decade earlier would have been promoted in terms of greater energy security and reduced energy costs are now largely framed by the authors in terms of reduced environmental impacts, and as cheaper ways of meeting emissions constraints; sometimes they are promoted to consumers on environmental grounds as well. On the supply side, environmental factors, from local planning issues through to regional and global emissions issues, are all seen as major determinants of the selection and development of technologies. The nature and implications of environmental concerns are examined further in the next chapter.

Another theme in all chapters, with the partial exception of gas turbines, is the sensitivity to government policy. Three particular aspects pervade the studies. One is that of energy pricing, including the extent to which governments seek to internalize the external impacts of fossil fuel activities. The second concerns the extent to which governments are prepared to intervene in cases where the market is clearly not likely to deliver optimal responses on the basis of price alone. Third is the form of utility regulation. As the nexus between production and delivery for both electricity and gas, and usually a natural monopoly, energy utilities play a central role and the way in which they are regulated can profoundly affect the outlook for competing technologies in both supply and demand.

There are however important differences in the underlying policy issues raised by the supply technology chapters as compared with those of demand. On the supply side, the studies seem broadly to accept the testing ground of the marketplace in the longer-term, *providing* it reflects the important issues and recognizes the high initial hurdles which new technologies face. There are significant concerns about the short time focus and high rates of return demanded by private investors as compared with the longer- term view of energy-environmental dilemmas; but this is not the primary issue. Cost is the ultimate arbiter: a level playing field (reflecting full social and environmental costs), fair access and open competition, and a helping hand over the initial obstacles which face new technologies aspiring to the big league, are the main things for which the studies ask.

The demand-side studies tell a very different story. The studies demonstrate, and some almost take it for granted, that many of the available efficiency improvements yield a greater economic return than almost any supply-side investments. In that sense they are economic on the criteria of the supply-side business - and even more so on the longer time horizons of society overall.[1] The problem is getting them in place. The demand-side studies all reveal similar obstacles, though in different proportions: widespread ignorance on the part of consumers; a lack of interest in operating energy costs relative to capital costs; and a variety of institutional and regulatory obstacles and distortions, sometimes resulting in perverse incentives. The studies show little or no faith that unmodified markets will exploit adequately the cost-effective opportunities; the calls are for governments to find some way of making people exploit what is undoubtedly good for society as a whole.

Another generic difference between the supply and demand studies concerns timescales. If the right levers are found, most of the identified options for improving the efficiency of demand-side technologies could be dominating sales by the end of the decade, with perhaps another

---

1. For a brief review and fuller discussion of the economic issues involved see M.Grubb, *Energy Policies and the Greenhouse Effect, Volume 1: Policy Appraisal*, Dartmouth, Aldershot, 1990, (Sections 3.2, 3.8 and 4.1).

decade for them to work through most of the stock. The analysts see little need to consider the impacts beyond this.

On the supply side, only combined cycle gas turbines are expected to have much impact on energy markets over a decade (though the building energy management study notes that small gas-fired engines for combined heat and power could also penetrate rapidly, with significant contributions to electricity supplies). That is only because they are already on the move and have already proved overwhelming advantages for many electricity systems of the 1990s; and even on the most radical growth scenario the potential penetration would still be far from exhausted, but rather, in the earlier stages of expected long-term global growth. Wind energy, similarly building on the kernels of existing operating experience, could be an international commercial industry with sales of thousands of megawatts and a turnover of billions of dollars by the turn of the decade; but this would still be in the early stages of exploitation compared with its potential total market more than a hundred times greater. On the same timescale, the most that clean coal technologies and photovoltaics could have would be a toehold in large-scale energy markets, and initial proof of commercial viability. The real promise of emerging supply technologies lies in the 21st Century.

12.2 Differences

This is not to suggest that either the demand or supply chapters tell uniform policy stories: each has its own characteristics. Vehicles are amongst the most complex because the industry is so big, because the car is such a 'potent symbol of success and status', and because the potential trade-offs against cost, power and even safety (both imaginary and real if pushed beyond a certain point) make mandated increases in efficiency highly contentious. More sophisticated approaches may be required, as discussed in the chapter, and worthwhile because of the scale of the industry.

For domestic appliances the issues are simpler; though in the extreme there would be trade-offs against cost and aspects such as appliance volume, substantial savings appear feasible with few if any significant trade-offs. The problem is more of indifference and ignorance. Clear and simple information, probably through mandated labelling given the failure of voluntary agreements, would be an important step where it has not already been taken. It is uncertain how much of the potential would be taken up in response to this, though research suggests that it would still leave the market far short of the theoretical potential.[2] Exploiting more of this potential would require more interventionist instruments such as standards.

New lighting technologies offer the largest proportionate savings, but buying them involves more than a trivial change for the purchaser: the key obstacle is that of the much higher initial cost, particularly for the replacement of filament lamps by compact fluorescent lamps, and the reluctance of consumers to balance this against longer-term savings. Overcoming this might involve special financing arrangements, perhaps through the utilities. Issues of compatibility with existing fittings, and characteristics such as the appearance and performance of the lights, are also important. For the service sector especially the education of fitters and consumers to encourage better design and use of lighting could be very significant. Continuing technology development, and perhaps the use of more extensive and aggressive marketing, will also affect the pace at which new lighting technologies are adopted.

Comparison of the appliance and lighting chapters reveals the importance of understanding technologies in detail if appropriate policies are to be adopted. On the surface, improving the efficiency of domestic appliances and of domestic lighting would seem to have a great deal in common as ways of improving the efficiency of electricity use in the home, and indeed end-use consumption statistics often put together appliance and lighting consumption. But in detail they differ greatly. Improving appliance efficiency involves a continuous range of very minor design changes with scarcely any visible impact in terms of capital cost or performance. Efficient lighting involves big step changes in the costs and characteristics of the technologies used, and the biggest energy

---

2. One study of the choices made by well-informed consumers in buying refrigerators which differed only in cost and efficiency concluded that only 40% of them bought the more efficient variety if it took more than 3 years to pay back (35% discount rate). Another 40% seemed to apply a discount rate above 60% (ie. requiring payback in well under 2 years). This contrasts with producer criteria of the order of 10-15% return, or around 7 to 10-year payback, while some public utilities have used 5%, (A.K.Meier and J.Whittier, 'Consumer discount rates implied by consumer purchases of energy efficient refrigerators', *Energy, the International Journal* 8(12), 1983). The fact that many of the consumers could have obtained credit to cover the extra costs and still make a profit is one indication that reality falls short of the economic ideal of rational consumers.

gains and fewest obstacles in terms of weight and performance will arise
from moving straight to the most advanced compact fluorescents with
integrated circuit control gear, when these are available. In the domestic
sector, saving electricity by improving appliance efficiency may be both
easier and have more impact than lighting changes (because appliances
consume about twice as much domestic electricity as lighting); but it is
the distinct high-technology image of compact fluorescent lights which
catches by far the greater publicity, rather than scarcely visible changes
to fridges or, indeed, to lighting in service buildings.

Finally, overall building energy management in the service sector raises
different issues again, arising from the divisions between builder, owner,
and occupier, the complexity of the options, and the extent of variations
between different buildings and building uses. There are a wide variety
of opportunities for improvements both in habits and technology: but
again, policy sophistication is called for. The larger service buildings
offer particular scope for addressing building energy use as an integrated
system, with education of building managers a key issue. For this, as for
service sector lighting, there is an important role for both the supplying
industries and for professional institutions, backed by government
support and demonstration schemes including the examples set by public
sector buildings.

On the supply side, the issues are again varied. Where adequate gas is
available, technical developments combined with regulatory and
environmental changes have in many areas already cleared the way for
rapid penetration of combined cycle gas turbines (CCGTs). Other gas
turbine technologies may require some support over the development
hurdle, either because the benefits they offer over combined cycles are
only marginal, or because they are best suited to markets which are
smaller (in the near term), less proven, and less familiar (as for biomass
gasifier turbines). Existing manufacturers may consider some of these
technologies, but the process would be greatly accelerated by additional
finance, from governments or possibly from unusual industrial and
international alliances combining supply and user interests.

The development of clean coal technologies is driven as much by
constraints on the traditional options as by government support, which
has been modest. It has not been aided by the embedded network of
electricity generation, manufacture and mining industries which have
often been reluctant to explore paths away from the tried and tested.

Given this, emissions constraints as well as government support for demonstration schemes are listed as important policy tools for encouraging the transition towards cleaner coal technologies, and progress is seen as intimately connected with the details of existing market conditions: the scope and requirements for repowering of existing coal stations; transfer deals with developing countries and World Bank finance; and the extent to which sometimes beleaguered mining industries can or will help to finance downstream developments. Progress may also depend on how utility regulation affects attitudes to risk and new ventures, and how it affects plant siting and planning procedures, including the potential and terms for using waste heat from power stations.

Wind energy is a different story: few realize that commercially it is now probably more advanced than most of the 'clean coal' technologies, largely because of the strong backing and experience in California and Denmark. But it is a technology with unusual characteristics by the standards of bulk power production, with units far smaller than conventionally used for bulk electricity, and it is driven largely by manufacturers historically independent of the electricity business. Thus wind had no natural industrial constituency within the energy business. Capital support for earlier developments was crucial, and support for private R&D remains valuable, as does 'common good' research such as wind surveys. But as the technology moves closer to market interests, other issues rise in importance: output credits to reflect the environmental benefits over fossil fuels, in lieu of environmental taxes; and calls for a level playing field in terms of system access and planning issues.

Solar photovoltaics (PV) shares some of these characteristics but is distinguished in many ways. PV is suited to an incremental and diverse approach, building steadily upon its successes in remote applications and in combination with small-scale diesel systems. PV is at a relatively early stage of development and only exists as a commercial industry because of these niche markets, which are steadily expanding. From this base the industry can probe a huge range of possibilities, from large arrays for meeting bulk air-conditioning demands to systems integrated into building surfaces. With an eye both to niche markets and the incidental benefits of being involved in them (such as greater visibility and activity in a wide range or markets), and the possible long-term central role of PV in global energy supplies, energy industries and governments have backed PV far more than other renewables. Fair access and pricing are important to the longer-term grid market, but because of the range of niche markets they are not the life and death issues which they are to wind energy. Industry itself is supporting incremental PV development; government finance is required for other areas. These include the more 'blue skies' options for radical improvements in the technology, and support for the special markets where PV is attractive, but which may be capital constrained. These are primarily the developing countries, so that issues of international technology transfer coupled with environmental constraints could be central to deployment.

12.3 Short and medium-term impacts

As noted above, one of the strongest distinctions between supply and demand technologies concerns the potential timescales of impacts. How much impact might the technologies discussed have on energy demand trends if heavily exploited? These questions can be illustrated with reference to the UK, on which many quantitative aspects of the case studies concentrate.

Realizing the potential savings suggested by the case study authors would mark a substantial departure from historical trends. Vehicle efficiency in the UK, in terms of realized on-road mpg, has been almost static since the mid-1970s, despite the sudden price rises of the oil shocks. As the study notes, in this case the technology has improved, but improvements have been offset by moves towards greater power and performance, and increased congestion. But considerable technical scope remains and the author argues that changes in policies could accelerate the pace of technical efficiency improvements, with gains of perhaps 20-40% in new vehicles over a decade or so, whilst the trends that offset this may slow and could be further discouraged by policy changes. Appliance efficiency has improved especially following the oil price shocks, but largely stagnated in the late 1980s, and the economic potential identified by the author would mark a significant acceleration even of the previous rate of improvement (especially for refrigerators), whilst trends to increasing appliance size (which offset some of the technical improvements) continue to slow. Lighting efficiency has also been relatively constant in the domestic sector, and still greater technical improvements would be available with a move to compact fluorescent lamps. Analysis is more complex in the service sector, where efficiency has been improving steadily, but the study still suggests that a significant acceleration of this trend would be possible through the adoption of better technologies. Still greater additional savings in the service sector (and light industry) would be available from the range of measures outlined in the building energy management chapter, in particular in conjunction with the adoption of BEMS. None of these options have yet had much impact on the UK building market.

The projected departures from historical trends, despite the sharp price rises in the 1970s for oil especially, incidentally support the view that they will not occur without significant non-price policy changes.

How much impact might such changes have overall on the UK energy market? To help address this question, Figure 12.1 shows the breakdown of energy use between fuels and sectors for the UK in 1990, together with the distribution of CO_2 emissions by end-use sector (including implied emissions from electricity and other energy transformation industries).

Figure 12.2 shows for the UK similar information for medium-term (notionally, the year 2005) sectoral projections. The methodology used is similar to that used for developing full bottom-up energy scenarios for the UK as described elsewhere,[3] and the assumptions in this projection are very similar to those described there for developing the 'business-as-usual' projections for the year 2000.

The relevant question here is not the accuracy of such projections, but rather the extent to which potential savings suggested by the authors would affect the overall picture. Table 12.1 shows the projected intensities of these end-uses in the 'business-as-usual' projection, and the levels which might be attained after 10-15 years if the potential for exploiting the demand-side case study technologies is heavily exploited. The hatched areas on Figure 12.2 then illustrate the impact which these selected changes would have on energy demand and CO_2 emissions.[4] Despite the limited range of options considered, the projected savings would make a substantial impact on the UK energy balance, reducing projected energy demand and CO_2 emissions by nearly 10%, sufficient

---

3. M.Grubb et al, *Energy Policies and the Greenhouse Effect, Volume II: Country Studies and Technical Options*, Dartmouth, Aldershot, 1991, Chapter 4.
4. Because the savings are greatest in electricity, the CO_2 savings are proportionately higher than the end-use energy savings. The generation mix is assumed to be constant, with 15% gas-fired generation.

Figure 12.1 Breakdown of UK energy use by fuel and sectorial CO₂ emissions, 1990

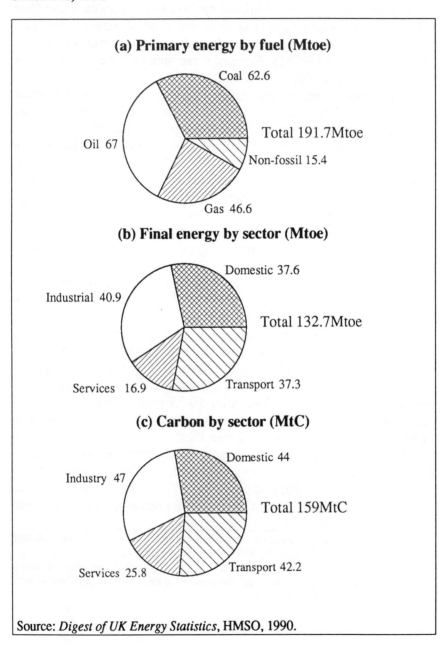

(a) Primary energy by fuel (Mtoe)

Coal 62.6
Oil 67
Total 191.7Mtoe
Non-fossil 15.4
Gas 46.6

(b) Final energy by sector (Mtoe)

Domestic 37.6
Industrial 40.9
Total 132.7Mtoe
Services 16.9
Transport 37.3

(c) Carbon by sector (MtC)

Domestic 44
Industry 47
Total 159MtC
Services 25.8
Transport 42.2

Source: *Digest of UK Energy Statistics*, HMSO, 1990.

Figure 12.2 Medium term sectorial projections for the UK with potential case study savings

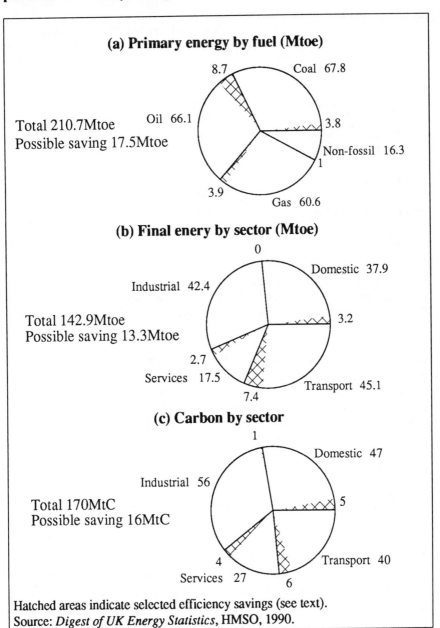

(a) Primary energy by fuel (Mtoe)

Coal 67.8

8.7

Oil 66.1

Total 210.7Mtoe
Possible saving 17.5Mtoe

3.8

Non-fossil 16.3

1

3.9

Gas 60.6

(b) Final enery by sector (Mtoe)

0

Domestic 37.9

Industrial 42.4

Total 142.9Mtoe
Possible saving 13.3Mtoe

3.2

2.7

Services 17.5

Transport 45.1

7.4

(c) Carbon by sector

1

Domestic 47

Industrial 56

Total 170MtC
Possible saving 16MtC

5

4

Transport 40

Services 27

6

Hatched areas indicate selected efficiency savings (see text).
Source: *Digest of UK Energy Statistics*, HMSO, 1990.

Table 12.1 Assumed energy intensities for case study technologies used to illustrate potential impacts

| | Energy intensity relative to 1990 | |
	Business-as-usual	Selected efficiency
Electricity		
Domestic refrigeration	0.9	0.6
Domestic lighting	0.9	0.5
Service sector[a] lighting	0.9	0.65
Service sector heating	0.9	0.7
Non-electric		
Cars	0.95	0.75
Service sector heating	0.9	0.7[b]
Domestic heating	0.9	0.8[c]

For a fuller explanation and details of projected usage levels, other assumptions and methodology, see M.Grubb et al, *Energy Policies and the Greenhouse Effect, Volume II: Country Studies and Technical Options*, Dartmouth, Aldershot, 1991, Chapter 4.

Notes: [a]service sector comprises all public and commercial premises other than industry. [b]Projected improvements corresponding to those discussed in the BEMS chapter. [c]Projected improvements corresponding to use of condensing gas boilers identified in BEMS chapter. Measures such as improved insulation (not selected as a case study technology) would increase the impact.

on this baseline roughly to stabilize primary energy consumption and slightly reduce CO_2 emissions.

Whilst the technologies studied cover some major end-uses and some of the biggest individual opportunities for savings, other important opportunities are not covered, including: non-refrigeration appliances; building insulation, micro-CHP and other opportunities for building energy management; large-scale CHP; non-car transport; and the whole area of industrial energy demand. The potential impact of efficiency improvements on demand therefore presumably exceeds those indicated in Figure 12.2 - a result entirely consistent with studies by other authors such as Skea[5] and Leach.[6]

The impact on fossil fuel consumption would be further enhanced by supply-side changes. On this timescale, the impact of clean coal and PV would be small, but that from gas turbines - displacing conventional coal stations with a 25-35% improvement in efficiency and a halving of CO_2 emissions, could be considerable. The impact of wind energy in the UK could also be becoming significant even on this timescale.

None of this is prediction: as emphasized by the case studies, it depends heavily on the policies discussed in the case studies and examined more broadly in the concluding chapter. In reality policy changes are likely to be slower, with extensive debate and experimentation. But the implications even of this small group of technologies for fossil fuel demand seem clear and too large to ignore.

12.4 Longer-term implications

Three great constants of energy history in the 20th Century have been those of increasing energy demand except under major energy price shocks, the dominance of fossil fuel steam technology for producing electricity, and the remorseless growth of oil use in transport. The case studies presented suggest that emerging technologies in developed economies could challenge at least the first two of these constants, whilst the third might at least be significantly curtailed.

Technology development is a continuing process on the demand side just as much as in supply. The improved technologies discussed in the case studies by no means approach physical limits. It has been estimated that the theoretical minimum energy required at the point of enduse to supply existing services in OECD countries is barely 5% of current primary energy consumption.[7] Although improving efficiency is likely to become steadily more difficult as the easier options are exploited first, the scope for further improvement is wide enough to suggest continuing potential impact on energy markets for many decades.

5. J.Skea, *A case study of the potential for reducing carbon dioxide emissions in the UK*, Science Policy Research Unit Special Report, University of Sussex, June 1990.
6. G.Leach and Z.Nowak, *Cutting carbon dioxide emissions from Poland and the United Kingdom*, Stockholm Environment Institute, Stockholm/London, 1990.
7. See discussion in M.Grubb et al, Volume Two, op. cit., Chapter 2.

On the supply side, the potential impact of the technologies considered is very large given enough time. Resources are not the critical constraint except on very long timescales, probably beyond the middle of the next century. As discussed in Chapter 2, gas resources accessible at competitive prices will probably last for several decades in most producing regions, and coal resources are far larger. Even for wind energy, for which the resources are generally more constrained than direct solar resources, the case study suggests that exploitable resources may be quite large enough to make it a substantial component (10-30%) of electricity supplies in many regions.

However, critical questions surround not only the actual markets, acceptability and economics of the different technologies, but also the rate at which they could penetrate, and the factors which will affect this. Energy supply technologies require very large investments, and particularly for those which require adaptation of whole systems, diffusion is clearly bound to be slow; as noted in Chapter 1, more than fifty years has been required for the major transitions from wood to coal, and coal to oil.

Such timescales would certainly apply to the development of a whole new transport infrastructure based on alternative fuels. The obstacles may be somewhat less for electricity supply technologies which feed into a system which is already in place; this is more closely aligned to the substitution processes discussed in Chapter 1, in which the bulk of system requirements for use of the product are already in place. As discussed there, new products under these conditions may grow at 20-30% a year. But the scale of the electricity industry is so great, and the starting point for most of the technologies is relatively so small, that even at these rates the implied timescales are long. For example, ambitious projections for the wind industry over the next forty years in Europe involve growth from current levels at about 25% a year for the next two decades, slowing to about 10% a year for the subsequent two decades - at the end of which, in 2030, wind would be providing an estimated 10% of European electricity supply.[8]

Nevertheless, gas turbines are already well on the move, and wind energy could be making a significant impact on new electricity generation especially in Europe and the US by the turn of the century. It

8. European Wind Energy Association, *Wind Energy in Europe - Time for Action*, EWEA, Oxford, 1991.

is conceivable that not long after the turn of the century, emerging supply technologies for electricity at least could be growing fast enough to offset any remaining upward pressures on energy demand in the industrialized world, forcing a reduction especially in the use of coal for power generation and perhaps overall consumption of oil and coal. If the pressures discussed in Chapter 2 continue and strengthen, the diffusion of all these technologies would accelerate. The end of the century could thus indeed mark the end of the era, spanning since the industrial revolution, in which economic development has been fuelled by rapid global growth in coal and oil consumption. But it will depend upon how the changing pressures evolve, and the response of policy and technology to this.

The Future Context

An underlying theme in most of the case studies is the interplay between economic and environmental forces. Future trends in economic policy and the extent to which environmental factors are reflected will help to determine the progress of emerging energy technologies and policy towards them. This chapter extends the discussion of Chapter 2 by considering the possible implications of economic and environmental trends for the future context of energy decisions, and the technological implications which this may have.

13.1 The economic context

It is almost a cliche to say that 'government has been in retreat' since the early 1980s. In many countries the old model of centralized public energy utilities, planning decades ahead in conjunction with government, is well in retreat. The history of government responses to the oil shocks and the changing pressures in the 1970s, as noted in Chapter 1, has not for the most part been a happy one: many have argued that free markets would have adapted better. Many (though by no means all) governments are wary of again taking a more active role in energy decision-making. The collapse of Soviet communism has been taken by many to reinforce more general principles of government non-intervention. Some specific illustrations of the way in which broad economic trends are influencing the energy sector have been noted in Chapter 2. They include reduced government backing for major technology development programmes, and attempts to open up utilities to more market-oriented financial pressures and competition. The latter includes changing the financial basis of utility planning, and opening up networks to independent suppliers, in some cases with 'common carrier' provisions which give large consumers the freedom to negotiate terms directly with suppliers, using the utility network as a transportation service. The privatization and division of the UK electricity system is the most striking example, but in many countries less drastic ways of opening up systems to competition are being pursued.

The implications of these trends for particular technologies are reflected in the case studies. To the extent that prices rise because of the higher rates of return demanded by private finance, and reduced government support, this may help to encourage greater end-use efficiency. More generally, the trends may help to make the energy sector more responsive in adapting to changing conditions and adopting new technologies in response to that. But the opportunity to fashion utility developments directly towards societal goals is reduced or gone, leaving governments with more indirect levers of altering the incentive structures for utilities. The crucial questions then concern the nature of incentives for the actors in these more open markets, and the kinds of technologies they may select.

In many respects, the trend towards a less active government role in the energy sector sits uneasily with much of the discussion in this book. Resource concerns remain as an important medium-term issue, with the knowledge that part of the problem in the 1970s was the failure to take anticipatory steps to limit dependence on OPEC, excepting the many eggs laid in the nuclear basket. However, whilst resource concerns are at least to some extent self-correcting through price rises (the main debates focusing on the unnecessarily damaging impact of sudden changes and the security implications of excess dependence), environmental concerns pose still more of a problem. Some issues of siting impact and accidents have become to an extent self-reinforcing, through the planning process and liability risks. But for other issues, such as most urban, regional and global pollution problems, government action is inescapable.

Encouraging technologies which help address such issues is not however just a matter of imposing environmental constraints, or incorporating environmental costs in energy prices. As the case studies emphasize a host of other factors are involved. More general technology policy has received significant attention from economists, and the notion that free markets (if adjusted for external costs) would deliver optimal responses has never received much support. Writing at the height of the

UK government's enthusiasm for free market policies, two government economists published a detailed study of the economics of technology policy.[1] Their extensive appraisal of the economic issues involved bears directly on the subject of this book and is worth quoting at length:

Technology policy should be seen as part of economic policy ... consisting of those economic policies specifically concerned with ensuring that firms, consumers, and government have access to appropriate and up-to-date technology at the lowest possible cost; with fostering invention and innovation; with encouraging the diffusion of innovations, new technologies, and technological best practice; and with ensuring that industry takes advantage of the economic opportunities offered ...

There is ... no single accepted way of classifying those instances of market failure which justifies government intervention ... however, for the purposes of technology policy the following taxonomy covers most of the relevant sets of circumstances:

(a) *Risk.* Firms, their managements, and their providers of capital may be averse to risk (and to uncertainty). This may cause them to under-invest in the development, appropriation and exploitation of new technology;

(b) *Information.* Markets can only allocate resources efficiently if participants are well informed about the opportunities open to them ... market mechanisms which might provide this information are inadequately developed;

(c) *Competition and market structure.* In some cases high up-front R&D costs may constitute a barrier to market entry. Where this results in inadequate competition, intervention to help firms overcome this barrier may be justified;

(d) *Externalities,* ie. where the actions of individual firms give rise to benefits which accrue to others which they cannot appropriate themselves, or costs which they are not obliged to bear;

(e) *Dynamic aspects of innovation and economic change.* There are circumstances in which the decisions of firms and of their suppliers

1. John Barber and Geoff White, 'Current policy practice and problems from a UK perspective', in Partha Dasgupta and Paul Stoneman, *Economic Policy and Technological Performance*, CUP, 1987.

of capital do not take account of the longer-term dynamic benefits which may result from a particular course of action ... [various examples of this are advanced, including 'infant industries', circumstances of rapid technological change, and applications where the technology could be of general strategic significance for the UK economy.] ...

The above list of market failures is not exhaustive, nor would it satisfy a fastidious theorist. Its purpose is to provide some practical guidance to help those involved in policy implementation to identify those circumstances in which government intervention may lead to net benefits to the national economy ...

The market failures listed here are exceedingly common: the case studies provide ample examples. In principle there is thus a rationale for potential government action in an enormous range of technological areas, far greater than actually experienced in most market economies - notwithstanding high-profile (and often expensive) support for some favoured technologies, action has been far less extensive than this list could in principle justify. The widespread retreat from active government policy towards energy technology in the 1980s has ironically been more a matter of practice than theory.

This retreat has partly reflected doubts about the practical effectiveness of such intervention in the energy sector, but also there has been limited incentive for governments actually to take corrective action. The key issue is therefore not so much whether support for emerging technologies can be justified in economic theory, but more whether government administration can enact the required policies effectively without incurring excessive hidden costs, and the extent to which policies would have to engage broad political and social ideologies about non-intervention - and whether the incentives to act are sufficient to overcome government reluctance and political opposition.

The oil shocks of the 1970s provided incentives to act, though the government response in many countries was dominated by supply-side R&D efforts. It is likely that concerns about resources and import dependence will one day again become sufficient to drive policy initiatives, but whatever the merits of measures to avert future difficulties, the incentive at times of falling energy prices has not appeared sufficient to maintain active energy technology policies. At the opening of the 1990s the driving force behind energy technology policy has become largely one of environmental concern, and a critical question for emerging energy technologies is the way in which environmental pressures will evolve and the implications they will have.

13.2 The environmental imperative

Environmental concerns are nothing new. Protests about the state of London streets and the river Thames date back to the Middle Ages, and energy developments have always been accompanied by an environmental shadow: mining conditions, oil spills, gas explosions, urban smog, hydro flooding, and nuclear fears, are not new stories. But as noted in the opening chapter, environmental concerns seem more pervasive than ever, and are affecting energy decisions at a more fundamental level than ever before.

Though concerns wax and wane, the underlying trend is upwards for clear reasons. First, human activities are steadily expanding: more people in bigger cities using more cars, travelling further, consuming more and generating more wastes. The underlying demands continue to grow and the impact will grow likewise unless technology responds and gets cleaner in all senses. The law of diminishing returns as applied to environmental clean-up suggests that the consequent efforts required also grow as the simpler options are used up.

Second, the environmental limit is not even fixed, but rather is itself tightening. This is in part because some finite environmental spaces are being used up (eg. waste deposits) but more because, as people get richer, there is less need to focus on immediate needs and they get more sensitive to the quality of their surroundings.

Finally, the cumulative effect now extends beyond the local and even regional level: the impacts of human activities have grown to planetary scales, and are changing the bulk characteristics of the surface and atmosphere, currently at an ever-increasing rate. And yet globally, the majority of people are in societies still at an early stage of industrialization, and population is still growing. With up to a ten-fold increase in global economic activity projected over the next century, global environmental pressures can only grow.

Ultimately, this reflects the need for *sustainable development*. Whilst there are various interpretations of the detailed meaning of this term, and about the extent to which sustainable energy development is possible with currently available technologies, it is implausible that the energy sector can remain immune from the pressures to move towards less environmentally damaging and more sustainable sources and technologies.

13.3 Technological implications

The case studies show some features of the responses to environmental pressures. One feature is the *interaction* of different environmental issues and the progression of responses. The clean coal study illustrates progression through three overlapping stages in power production. The first is 'good housekeeping': coal washing, good maintenance to get burning as clean as possible, high stacks to prevent pollution accumulating in one location. The second is a succession of 'end-of-pipe' measures: precipitators to remove particulates, followed by flue-gas desulphurization. But end-of-pipe fixes carry their own limitations, notably those of generating other wastes, costs, and the energy penalties involved which lower overall efficiencies. Eventually the cumulative weight of these pressures on a technology which had already been taken close to its physical limits have brought steam coal plant to the end of the road, too costly and inefficient to compete when the cleaner gas option opens up. New coal plant in such a context is likely only to be viable by moving to a third stage by making fundamental changes in the underlying technology, using new and inherently cleaner processes. These clean coal technologies still need development, but in the long run may well prove not only cleaner but more economical overall than traditional approaches, though environmental concern has been the driving force behind the technological leap.

Car exhausts have followed a similar pattern in the first two stages. Better housekeeping in terms of improved engine combustion and control, and fuel processing (eg. non-lead fuels) could meet earlier environmental requirements. End-of-pipe solutions included various exhaust and fuel clean-up procedures, culminating in the three-way catalytic converter and reformulated gasoline. But these do not come without trade-offs: energy penalties at the refinery and perhaps in cars, added capital costs, and increased requirements for special materials and wastes from the converters themselves. The vehicle case study is optimistic that the room for further efficiency improvements is substantial, and more than enough to offset the costs of clean-up measures so far adopted. It therefore remains an open question whether or when transport will start a shift to a third stage of responses to radically different transport technologies, but as noted, some believe this to be not only feasible but not so far away, perhaps drawing on technologies and lessons emerging from the cradle of the measures taken to combat pollution in the Los Angeles Basin.

The addition of end-of-pipe clean-up generally resolves more pressing environmental problems at the expense of others, in part because of the overall reduced efficiency. Deeper process changes, improved end-use efficiency and moves to alternative fuels can address a wider range. To date, such moves have mostly reflected specific non-CO_2 constraints combined with general economic pressures. Over the next decade or two, the scope for continuing to clean up fossil fuels, especially when combined with the arrival of the cleaner gas, suggests that non-CO_2 environmental pressures need not threaten the dominance of the hydrocarbon economy. In the longer term, the probable limitations even of clean-coal technologies in terms of costs and residual wastes, combined with resource concerns about over-dependence on foreign gas and oil, points to the unavoidability of measures to exploit higher efficiency and some non-fossil technologies.

In practice several factors serve to bring forward the conscious promotion of greater efficiency and non-fossil sources. The first is simply foresight of the above constraints under 'business-as-usual' fossil fuel consumption. The OPEC price shocks and the damage to European lakes and forests have imprinted lessons about the costs of waiting until problems are inescapable before attempting to address them. These are but forerunners of what could happen as the progressive depletion of resources delivers effective control over remaining reserves to successively more limited producing regions, and as environmental impacts, possessing immense inertia on the global scale, accumulate.

The second factor is more explicit concern about the greenhouse effect. The uncertainties are many but it seems implausible that human interference with the radiative balance of the atmosphere, one of the most basic of the planetary regulatory mechanisms, can continue and accelerate without adverse consequences. International negotiations are proceeding and the political pressures for some kind of action are considerable. Furthermore, if measured but serious steps to limit the accumulation of CO_2 are not taken in the near future, the chances are high that a combination of climatic extremes and more scientific consensus will engender a panic reaction, which could have profound and costly consequences for energy production and broader society.

Third is the recognition of the growing technical possibilities and the benefits they offer. There is increasing appreciation, backed by most of the case studies in this volume, that measures to improve efficiency in both supply and demand need not be costly, and could indeed be economically beneficial. Even the non-fossil options hold considerable economic promise, especially when the general external costs of fossil fuels, and the likely trend of further advances, are taken into account.

When coupled with a recognition of just how long energy technologies take to emerge to market maturity, early starts are valuable. Hence the broad interest in the promotion and acceleration of the full range of cleaner technologies and options for improving efficiency in both supply and demand. The interplay between economic and environmental forces will be an enduring theme in energy, and technologies which reduce environmental impacts and aid sustainability will inevitably assume a greater role. Yet as the case studies emphasize, the nature and rate of progress of such technologies will depend heavily on how energy policy evolves. This book closes with a review of the policy lessons and implications.

It is not only technologies which evolve in the energy world. Policies too are in a continual state of evolution, reflecting changing pressures and goals, the lessons learned (and sometimes forgotten) from previous experience, and the specific needs of both existing and emerging technologies. This final chapter does not attempt to analyze in depth the economic and political issues raised by different policy tools, as has been presented by one of the editors elsewhere for some of the policies involved.[1] Rather, this chapter seeks to draw together and interpret the various policy issues raised by the case studies, as a possible pointer to the kinds of policies which may emerge hand in hand with emerging technologies themselves.

14.1 Research and development

With the exception of coal and photovoltaics, the role of explicit research and development efforts does not feature prominently in the case studies. This in part is a reflection of the focus on 'emerging' technologies, which by the criteria explained in Chapter 3 are chosen to avoid 'blue skies' options which clearly depend upon major technological advances. Nevertheless, the studies reveal important points about the nature and role of past R&D, and its potential impact in the future.

The lessons from past direct government R&D are not heartening. As noted in Chapter 1, major public programmes to develop particular technologies, such as the nuclear and synfuels programmes, have not generally yielded the success expected. Figure 14.1, which illustrates the total direct R&D expenditure by government members of the

1. M.Grubb, *Energy Policies and the Greenhouse Effect, Volume 1: Policy Appraisal*, Dartmouth, Aldershot, 1990.

Figure 14.1 Energy R&D expenditure by IEA governments, 1988-1990

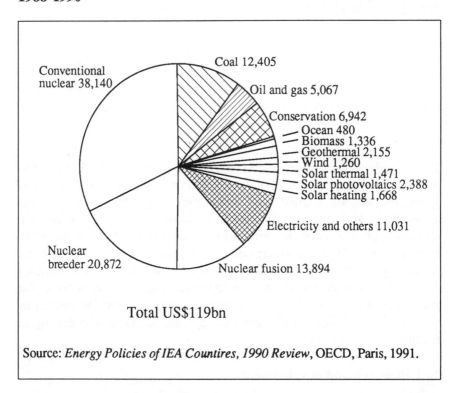

Source: *Energy Policies of IEA Countires, 1990 Review*, OECD, Paris, 1991.

International Energy Agency since 1977, suggests a more general misdirection of R&D expenditure. Whilst the limited scope of this study means that no direct correlation between such expenditure and the technologies examined can be expected, nevertheless it is striking that there appears to be no discernible relationship between the allocation of government energy R&D expenditure over the past fifteen years and the technologies which now appear most likely to have a significant impact on energy supply and demand over the coming decades.

It is more difficult to disentangle the scale and impact of private R&D initiatives. However, the previous chapter noted that most of the technologies considered reflected primarily a process of 'demand pull' rather than 'supply push'. Substantial private R&D resources have been applied to improving existing production technologies, such as oil exploration and extraction, steam turbines, and gas turbines especially for aircraft. But turbine manufacturers seemed very slow to exploit the long-visible potential for combined cycle systems, and they are not putting significant resources into other advanced cycles (as discussed in the gas turbine chapter). Despite the immense scale of the national and international coal business and the pivotal role of coal in most electric utilities, R&D expenditure on clean coal technologies by either has been modest. Wind energy has been largely developed by independent companies, with occasional backing of large construction and aerospace companies.

Thus it appears most industries have been reluctant to pursue R&D efforts outside their main area of existing business unless conditions force them to do so, the primary exception being the efforts by oil companies to develop photovoltaic cells. For the demand-side technologies as well, the resources devoted to developing and promoting more efficient technologies in the absence of direct pressure to do so have been limited.

The most effective R&D on new energy technologies appears to have been that carried out primarily by private companies in response to direct pressures or strong government incentives. The most striking example is the history of wind energy over the 1980s, but the vehicles chapter also illustrates the rapid response of car manufacturers to the US CAFE standards, whilst the recent development of clean coal technologies owes much to government backing of private R&D reinforced by environmental legislation which was clearly tending to squeeze out existing coal technologies.

Government attitudes towards R&D support vary over time, between countries, and between different technologies; many governments have in fact continued or extended general support for companies working in leading-edge technologies in the belief that this is important for international competitiveness.[2] But, as enthusiasm for direct government R&D has declined, and notwithstanding the examples given in case studies, energy has seen relatively little of such support for private

---

2. One writer suggested that 'two major convictions have gained considerable strength ... it is widely believed that economic growth can only be regained and sustained with the aid of a high rate of technological innovation ... and [that this] cannot be achieved without government support'. (E.Braun, *Wayward Technology*, Frances Pinter, London, 1984). See also discussion in previous chapter on broad technology policy.

innovation. Evolving pressures may lead to a more widespread extension of such support to emerging energy technologies.

The extent to which explicit government support and direction is needed, and the appropriate mechanisms, vary. For the big and highly competitive car industry, R&D expenditure is in any case a central part of the business and what matters is how incentives affect its direction, which will depend primarily on the structure of producer standards and consumer incentives. In this respect fridges may have much in common with cars, especially concerning the replacement of CFCs, where manufacturers have to find replacements for CFCs, but without directed support they are likely to follow the path of least resistance rather than directing efforts at the most efficient options.

Developments in the mainstream of established industries (eg. gas turbines (now including CCGTs) and refinements to existing wind turbine systems) will continue irrespective of government policy. Outside the mainstream, companies are much more cautious and developments may depend on external support. The gas turbine study suggests that this may be required to encourage the industries to take on the risks associated with development of more advanced gas turbine systems for alternative fuels, whilst development and exploration of novel approaches in PV, wind and clean coal would clearly be hastened by greater R&D support from government or, potentially, from external industrial sources such as utilities - as illustrated for example by the pivotal role of the major Californian utilities in developing new wind and photovoltaic systems.[3]

Whilst it is easier to focus on supply technologies, the role of R&D on the demand side is also important, for technologies ranging from advanced vehicle design to integrated circuit fluorescent lamps and the various options described in the case study on building energy management systems. Fulkerson et al[4] note a wide range of demand-side research needs. But along with explicit support, the focus of private R&D on demand technologies is inseparable from the impact

---

3. Notably, the joint ventures between Pacific Gas and Electric and US Windpower to develop the variable-speed wind turbine, and between Southern California Edison and Texas Instruments to develop new PV modules.
4. W.Fulkerson et al, *Energy Technology R&D: What Could Make a Difference?*, ORNL-6541, Oak Ridge National Laboratory, Oak Ridge, Tennessee, December 1989, (Volume II: End-use Technologies).

of more general policies to promote the take-up of more efficient technologies, and hence to create markets for products which may in fact require modest R&D outlay. The same is true, but to a lesser extent, for supply technologies. The rest of this chapter summarizes the policy options involved not in developing new technologies, but in encouraging their application.

14.2 Incorporating external costs

Chapter 2 noted that in addressing environmental concerns, there has been a trend away from using technology-based controls and regulations towards more flexible 'umbrella' controls on emissions levels. Mandated emissions targets allow firms flexibility in the technology used to meet the environmental goals set. In some cases, particularly in the US, the flexibility has been further extended by the use of tradeable emissions permits which allow companies to trade obligations whilst ensuring the overall target is met. The other more market-based route is to tax pollution in a more explicit effort to 'internalize' the external costs. Ultimately, the net impact on the economics of competing technologies should be similar, since the price of tradeable permits should rise to the level of the tax required to constrain emissions by the same amount.

The impact depends upon this price, which in turn should reflect the environmental or other external costs involved. As discussed in Chapter 2, the external costs associated with current energy production are very uncertain, particularly with respect to pollution damage, but estimates suggest the figures may well be high enough to make a considerable difference to the relative costs of different options. The wind and PV case studies especially identify the inclusion of external costs as a key element in making these technologies competitive on a wide scale, and obviously price rises - and the perception that prices may rise because of moves to include external costs - will enhance the uptake of efficient technologies.

Some governments are seeking to develop tax and/or credit schemes which are intended to reflect such external costs, but few governments have yet put the principles into practice, and attempts to do so have met with strong opposition. A few governments have introduced pollutant taxes, but these have frequently been more to do with raising revenue for pollution-abatement than directly trying to alter behaviour or technology. Some north European governments have introduced a carbon tax, and the European Commission has now advanced detailed proposals for a combined energy/carbon tax as part of measures to meet the EC's goal of stabilizing CO_2 emissions. Initial response has indicated that it will not face an easy passage, and in general it seems that for many years to come, relying primarily on pollution taxes as an instrument of energy-environmental policy will be more a matter of economic ideals than of implemented policy. If correct this will place even more onus on other policy approaches.

14.3 Supply-side technology policies

On the supply side, this book has not addressed technologies involved in the production of fossil fuels. Technologies such as those listed in Chapter 3 could have an important impact on available reserves and costs. But they already form part of the ongoing development of the mainstream fossil fuel production business, so changing pressures outside the production business have relatively little impact, and the converse is also true. To the extent that governments may wish to stimulate such technologies, the main instrument is probably straightforward corporation tax incentives.

The supply-side studies have focused on the electricity sector, where the choice of technology is wide and sensitive to economic and environmental pressures. In addition to 'getting the price right', issues of fair access to systems and initial support to help at minimum correct for externalities and overcome entry barriers, or more extended support to speed up the slow process of diffusion, are very important.

Of the technologies considered, trends towards more liberalized and competitive electricity sectors seem of unambiguous benefit only for combined cycle gas turbines, though more open access may also be an important component of the required package for encouraging wind energy and photovoltaics (and possibly also small-scale CHP in its role as a part of building energy management in large service buildings). The changes do not favour coal technologies, new and old alike. For wind and PV, these reforms are double-edged. They can remove some of the institutional obstacles to novel supply sources, and private finance favours the short lead-times and low risks involved - but not the capital-intensive nature of these technologies.

In fact, in largely private-financed electricity systems, it is hard to see any electricity technologies competing unsupported against CCGTs in areas with access to adequate cheap natural gas, perhaps even if environmental externalities are reflected in the costs. If changes in utility ownership and the consequent dash for gas continues, this may again turn the energy agenda towards issues of diversity and security of supply, especially where this coincides with environmental or social goals (as for the renewables and often clean coal technologies), as a rationale for supporting such sources. This would imply support at least to the point where they form a few per cent of supplies and can form an industry capable of rapid expansion - which means more than just backing for R&D.

The key issues are then likely to be not just the degree of support, but its appropriateness, as illustrated particularly by the wind chapter. The most effective support is likely to focus on minimizing obstacles (such as by streamlining planning procedures, and overseeing terms of access for small-scale power producer); credits for the power produced (to reward performance, rather than just capacity installed); and 'common good' support which is of benefit to the whole industry. Fundamental research and development on the technology is one kind of common good support, but it also can comprise ancillary knowledge such as wind atlases and experience in trying to integrate photovoltaics into building surfaces and building energy management.

14.4 Promoting energy efficient technologies

In many respects the demand side is more complex still. The studies argue strongly that only a fraction of the economic potential will be taken up without a more active and conscious government role. 'Lubricating the market' with public information has a role, but not a dominant one. Giving clear information a higher and unavoidable profile, as with mandatory appliance labelling, would push this further. Both might be considered within the remit of 'free market' governments. But this alone would still appear to leave considerable economic potential untapped.

This might matter less if there were not such deep concerns about energy provision, but when faced with the dangers of growing import dependence and the need to meet CO_2 and other emissions goals, the problems are more serious. The most extensive ways of exploiting the technologies available involve all the things from which governments have been retreating; ways of telling people what to do, which technologies to select, which standards to meet.

In some cases blunt instruments may be appropriate. In many countries, building standards already determine minimum levels of insulation. A good case can be made for domestic appliance standards, in part because the trade-offs against performance and even capital cost appear to be so slight, and in part because the relatively low cost and universality of appliances means that the information costs of seeking out more efficient appliances may be significant relative to total expenditure. These information costs are saved by having national standards.

But a more general answer probably lies in greater policy sophistication. In some cases, a principle of using standards which leave firms and individuals maximum room to manoeuvre within them could be exploited: applied to certain classes of service sector buildings, this would encourage exploration of options spanning insulation, more efficient lighting, CHP, and energy management education and systems. Tradeable vehicle efficiency credits would also fall into this category, allowing firms to specialize in particular sectors of the market without distortion, whilst ensuring that the whole fleet meets given efficiency goals.

Other options could be to amplify existing but weak incentives to consumers. Again, vehicles provide an example, for example with the 'gas guzzler' tax and 'guzzler to sipper' rebates. This is one particular form of cost-transfer incentives, transferring part of the operational costs and savings to the capital, where consumers weight them much more strongly.

14.5 Utility regulation and demand-side management

The importance of utility regulation has been highlighted on many occasions. In most developed countries, at least half of primary energy consumption is channelled through electric and gas networks which are natural monopolies. The trends towards more open access, private participation and competition address only one end of the system. The other end of the wire, or pipe, has received far less attention in most parts of the world. The technology used to meter use is antiquated compared with the possibilities, which include systems for time-of-use pricing and the provision of 'interruptible' loads at lower costs. Better metering technology is available that could itself help to give consumers a better understanding of the consumption of equipment in their homes and offices, such as appliances and lighting.

Changes in utility regulation can be used more explicitly to promote greater efficiency. 'Least-cost planning' for utilities, and the more general issue of utility demand-side management and energy efficiency (DSM/EE), is highlighted in all the building-oriented demand studies. These have become generic terms for regulatory systems which involve the utility in actively promoting greater end-use efficiency (it is usually considered with respect to electricity but is also relevant to gas).[5]

This can be done by requiring utilities by law to compare generating options against end-use options and invest in whichever is the cheaper. But since this requires utilities to make investments that lose them money (by reducing sales unless recovered by other means), it raises complex questions about financing and compensation, as well as awkward monitoring issues given that the interest of the utility is then to invest in the least effective end-use programmes that they can get away with. Alternatively, utilities can be given direct incentives to explore end-use options. The regulatory issues are complex, and vary according to the structure of the utility: least-cost planning 'is not a single fixed system, but a continually evolving regulatory framework and set of instruments which differ according to the situation of different utilities'.[6]

There are now many examples from utilities in the US. As one example, the California Public Utilities Commission (CPUC) adopted new regulations in 1990 which allows the investor-owned electric and gas utilities in California not only to recover the full cost of their demand-side management programmes - including the recovery of any lost revenues due to reduced sales - but actually profit from these investments. Californian utilities can, for example, keep 15% of the estimated savings resulting from their energy efficiency investments as profits for shareholders.

This has stimulated a range of programmes. To promote efficient appliances, in 1991 the Southern California Edison utility arranged payments to warehouses that stock appliances that exceed the (already relatively stringent) Californian efficiency standards, and to retailers that display and sell them. Other programmes involve direct visits by utility staff to homes, service buildings and industry to advise on more efficient equipment, and in some cases to install it. As predicted by the case study

---

5. Ian Brown, *Least-Cost Planning in the Gas Industry*, Office of Gas Supply, London, 1990.
6. M.Grubb, Energy Policies and the Greenhouse Effect, Volume 1, op.cit, p.136.

authors, efficient appliances, lighting, and overall building energy management feature strongly given such incentives.[7]

The utility interest is simple: usually the shareholders make money from such initiatives. The impact of these regulatory changes in the US is profound:[8]

> The electric power industry's gross investments in DSM/EE are estimated to total $2bn in 1991, and are projected to reach $5bn by 1995, and perhaps as high as $10bn by 2000. These programs are expected to reduce the growth of peak demand and energy consumption. Some utilities expect to meet 50 to 100% of their additional resource needs in the 1990s from DSM and end-use EE improvements ...

Least cost planning arose originally because of difficulties in getting planning permission for new power stations when saving an equivalent amount of energy through DSM was demonstrably cheaper. But now that it has become an established part of utility practice in some areas in the US, the examples seem likely to spread rapidly as a way of promoting the uptake of more efficiency technologies for helping to address the broader range of energy and environmental concerns.

14.6 Technology transfer and developing countries

International concerns add another dimension. Globalization of business together with growing economic and environmental interdependence was another theme highlighted in Chapter 2. This has many faces.

One is the specific issue of the technology choice of poorer countries as they develop, and the role of technology transfer within this. About four-fifths of the global population live in developing countries, and that proportion is growing as populations expand. The global environmental and resource implications of growth following in the tracks of the western model would be immense, and could not be sustained. Furthermore, the costs of moving towards more sustainable patterns of energy development are generally much lower if the changes are made

7. Fereidoon Sioshansi (Southern California Edison), Lecture to British Institution of Energy Economics, Chatham House, September 1991.
8. F.Sioshansi, 'Recent US Developments in DSM/EE and their long-term implications: an international perspective', 15th Annual International Conference of the International Association for Energy Economics, Tours, France, IAEE, forthcoming, May 1992.

whilst the basic infrastructure is still being developed. Consequently there is widespread concern that developing countries should choose much more efficient products and environmentally cleaner technologies as they develop: the cream of emerging technologies.

The reality of course is frequently the opposite. The developing countries have other priorities and are desperately short of capital. Technologies are generally selected not on grounds of lifetime costs and minimal environmental impact, but on grounds of whatever is available at lowest first cost. Even where improved efficiency in the energy system would make general economic sense it can be desperately difficult to implement. Prices are often subsidized. Price reform (including removal of subsidies) is difficult enough in any society; especially when people already cannot afford basic fuels it is political dynamite. Most efforts are focused on patching up what is available and trying to minimize supply shortfalls, rather than any longer-term supply and demand planning. The gap between the ideal of transferring the most advanced technologies and the reality is painfully large.[9]

Even trying to discourage the transfer of old and inefficient technology is fraught with difficulty. Trying to prevent the transfer of second-hand vehicles, for example, could be highly contentious, for the net effect could be to deprive the poor in developing countries of the only vehicles they can afford, and deprive dealers of the income from selling them. The same logic would apply to other technologies. The vehicles chapter suggests that more sophisticated incentive schemes such as the guzzler-to-sipper rebates discussed for new vehicles might be applicable in some form. For a variety of economic and political reasons, much will hinge on the extent to which developing countries themselves seek to promote emerging technologies in domestic production, particularly of end-use technologies such as refrigeration.

If there are to be responsible efforts to enhance the transfer of more advanced technologies, the opportunities are large and the case studies illustrate this. For more efficient vehicles and appliances especially, the 'leverage' gained by directing some resources towards encouraging more efficient technologies would be immense. On the supply side, there is a

---

9. For a graphic account of the complexities and contradictions facing more sustainable paths for energy development in a developing country, see A.Mathur, 'India: Vast Opportunities and Constraints', in M.Grubb et al, *Energy Policies and the Greenhouse Effect, Volume II: Country Studies and Technical Options*, Dartmouth, Aldershot, 1991.

greater obstacle of capital cost. Nevertheless, gas turbines and support for gas infrastructure development, clean coal, wind energy and photovoltaics could all feature strongly; as the cost falls, PV especially seems very well suited to conditions with strong sun, isolated applications or weak grids, and limited capacity for managing and maintaining more complex systems. The World Bank's $1.2bn Global Environmental Facility is in part set up for such purposes. In all, the topic of international technology transfer linked to goals of more sustainable development is both highly politicized and very complex, requiring a range of innovative developments.[10] The issue is its early days; but the ways in which these issues are resolved will have profound consequences for the global role of emerging technologies and broader energy issues in the 21st Century.

14.7 Globalization and competitiveness

The continuing globalization of business will also affect the international progress and implications of emerging technologies. To the extent that companies seek to adopt similar standards and processes in different regimes to minimize the obstacles to international transfers, may encourage the more rapid spread of better processes. But the development of policy in response to the evolving pressures in the energy business, and the response of emerging technologies to these, is likely to raise highly contentious issues of competitiveness.

This will be one of many issues affecting technology transfers to developing countries, where local industries frequently may not meet the standards of industrialized country competitors. The issue could be equally potent within the industrialized world. The study of vehicle efficiency noted the dissent generated in the US by the CAFE standards, which were held to favour the smaller and more efficient Japanese vehicles. The appliance study hints strongly at the potential for similar tensions in the UK over appliance standards. There is a natural tendency for governments to try and protect domestic industry, but extensive

10. 'The need for innovation on all fronts is paramount. Neither policy reform, nor management improvement, nor technological innovation, nor financial innovation alone can solve the problem of increasing the per-capita delivery of electrical services tenfold under capital and environmental constraints. A multifaced, integrated approach will be required, posing difficult challenges for the international community' (D.Jhirad, 'Power Sector Innovation in Developing Countries: Implementing Multifaceted Solutions', *Annual Review of Energy 1990*, Vol.15 pp.365-98).

studies concur that despite the best efforts of government and industry, '... technology simply does diffuse globally. The question is then how much one can abet or retard it'.[11]

The case studies do not paint an enviable picture for the UK, which usefully illustrates key industrial issues of wider import. UK models do not feature significantly in the list of efficient appliances, nor of high efficiency prototype vehicles. The appliance case study notes that partly because of pressure from its industries, the UK has opposed most European initiatives for encouraging more efficient appliances, even that of mandatory labelling to provide consumers with information on appliance energy consumption. A reluctance to support emerging supply-side technologies has also left the UK lagging in this, as painted particularly starkly by the case of clean coal technologies, and also by the precarious state of the UK wind industry facing Danish competition. Environmental needs and resource concerns form the main plank for those arguing in favour of greater measures to support emerging technologies, arguments which tend to be countered by concern expressed about the impact on established UK industry. Yet even on straightforward industrial grounds, resisting such pressures is only advisable if the UK can continue to hold out against tougher environmental and efficiency standards.

This brings us back squarely to the issue of prediction with which this book opened. The energy business has lost billions of pounds through not foreseeing future trends adequately. UK car manufacturers may have lost billions in failing to foresee the triumph of catalytic convertors in Europe. Many other examples exist of the reluctance to recognize changing forces in the energy world, or of incorrect beliefs that they could be resisted. In parallel, there have been failures by industry to recognize the potential for new technologies to address rising concerns, and of government to reflect the pressures in legislation which could move industry in this direction at an earlier stage. The appliance study suggests that in resisting any incentives for UK industry to move towards more efficient appliances, even that of labelling, the old story will be

---

11. Jesse H. Ausubel, 'Rat Race Dynamics and Crazy Companies: The Diffusion of Technologies and Social Behaviour', in Nebojsa Nakiacenovic and Arnulf Grubler (eds), *Diffusion of Technologies and Social Behaviour*, Springer-Verlag, Berlin/Heidelberg, 1991, p.13.

rerun again when the UK finally and inevitably loses the battle against efficiency standards or other strong incentives in the EC.

The UK provides a particularly clear case because the evolving pressures in the energy business are combined with the trend towards the Single Energy Market and broader European economic union. But globally, many of the same forces are at play, albeit on a slower timescale. Japan, California and Denmark all provide contrasting images, as regions which have responded to domestic environmental and economic pressures with a range of varied measures to promote environmental quality and greater energy efficiency. These policies have sometimes incurred costs and illustrated pitfalls, but have also positioned these regions well with respect to emerging international markets that increasingly have to respond to the same pressures. The most successful industries and countries will be those which can learn from past mistakes, but also anticipate future needs, and thus develop prescient and effective policies to promote emerging energy technologies.